Inventing For Dummies®

Characteristics of Successful Inventors

To be a successful inventor, it helps if you are:

- Able to tolerate failure
- Achievement-oriented
- Competitive
- Creative
- Demanding
- Goal-oriented
- Highly energetic
- Independent
- Innovative
- Inquisitive
- Open to feedback
- Persistent
- A risk-taker
- Self-confident
- Self-motivated

Sources for Innovative Ideas

- Demographic changes in society
- Imagination
- Luck
- Energy-saving
- Problem-solving
- Health and Safety
- Dissatisfied customers
- Vision

A Typical Product Lifecycle

Most inventions are in and out of the market within a few years. Years one to two and a half are the introductory years. You take your product through the idea and development stage while you check out the commercial market, apply for legal protection, and try to obtain funds. At about year two and a half, the product enters into the market and sales start. They peak at about year five. During year six, you must cope with new competition and competitors and by year seven, the sales are in decline. Remember, although your patent life is for as much as 20 years, the market life is quite different and nearly always shorter.

Golden Rules

- One inventor can change the world.
- He who has the technology, wins.
- Getting a patent doesn't guarantee that you're going to make money – less than 5 per cent of patented products ever do. A patent is only one piece of the total product pie.
- Many successful inventors don't invent for the money, they invent because that is what they do.

Going to Market

You have three choices on how to bring your product to market:

- You do everything – manufacture, market, and sell.
- You subcontract the manufacturing and concentrate on marketing and selling your invention.
- You license your intellectual property rights to a company that arranges the manufacturing, marketing and selling aspects and pays you a lump sum and/or a royalty (a percentage on each unit sold).

Protecting Your Intellectual Property

Protect your idea with a patent, registered design and/or registered trade mark. Be aware of and make use of your automatic rights such as copyright. If your invention is really good, you are more than likely going to have to defend your rights. Get good legal advisers, as a product is only as good as it can be defended in a court of law. Use these tips:

- Keep good records of your concept and its development.
- Conduct a patent search.
- Get sound legal advice from a professional.
- Use confidentiality, non-disclosure, and employment agreements with anyone and everyone with knowledge of your idea.

For Dummies: Bestselling Book Series for Beginners

Inventing For Dummies®

Cheat Sheet

Planning Your Marketing Strategy

To paraphrase Thomas Edison, 'I'm not going to invent anything unless it will sell'. You can create the greatest invention in the world, but unless the world knows about it, what difference does it make? The way to make a difference is through creative marketing and advertising efforts.

Some marketing tips:

- Find out as much about the industry your product sells in as you can. Educate yourself about wholesalers and distributors, manufacturers, and competition.
- Use every resource you have for assistance, including the Internet, the British Library, the UK Intellectual Property Office, Business Link, local college and university business schools, and inventor organisations.
- Let the customer drive your product. Find out what customers like and – even more importantly – *don't* like about your invention and try to accommodate their preferences.
- Likewise, realise that the customer determines the price of your product. You charge what the customer will pay and work out your profit margins from that.
- Most importantly, once you have customers, pay attention to them.

Licensing Tips

Only a small percentage of all patents are licensed, but it can be a profitable way to go if you can do it. Use these tips:

- Make contacts at trade shows and elsewhere who can help you get a foot in the door of the right companies and help you avoid the wrong ones.
- Take time to develop negotiating skills.
- Focus and concentrate on obtaining a win-win licensing agreement.
- Know when to sign on the dotted line and when to walk away.
- Leave room for further rounds at the negotiating table by planning your strategy and tactics, and by listening and studying.

Money Matters

Make certain you have adequate financial resources. You don't want to start developing your product and then be forced to quit halfway through. Funding sources can include:

- Anyone and everyone who tells you that your product is wonderful. Ask them to put their money where their mouth is. Don't deny them a wonderful investment opportunity.
- Investor and entrepreneurial networking meetings. These can be really helpful. Follow up on any referrals; nearly all funding comes from them.

If you have to bring investors into your business, make sure to look for more than just money. You want an investor with business savvy, experience, and funding friends.

For Dummies: Bestselling Book Series for Beginners

Inventing
FOR
DUMMIES®

by Peter Jackson, Philip Robinson,
and Pamela Riddle Bird

Foreword by Trevor Baylis, OBE
Founder of Trevor Baylis Brands

John Wiley & Sons, Ltd

Inventing For Dummies®

Published by
John Wiley & Sons, Ltd
The Atrium
Southern Gate
Chichester
West Sussex
PO19 8SQ
England

E-mail (for orders and customer service enquires): cs-books@wiley.co.uk

Visit our Home Page on www.wiley.com

Copyright © 2008 John Wiley & Sons, Ltd, Chichester, West Sussex, England

Published by John Wiley & Sons, Ltd, Chichester, West Sussex

For general information on our other products and services, please contact our Customer Care Department within the U.S. at 800-762-2974, outside the U.S. at 317-572-3993, or fax 317-572-4002.

For technical support, please visit www.wiley.com/techsupport.

Wiley also publishes its books in a variety of electronic formats. Some content that appears in print may not be available in electronic books.

British Library Cataloguing in Publication Data: A catalogue record for this book is available from the British Library

ISBN: 978-0-470-51996-7

Printed and bound in Great Britain by Bell & Bain Ltd., Glasgow

10 9 8 7 6 5 4 3 2 1

WILEY

About the Authors

Peter Jackson has a long history of involvement in protecting new ideas and converting them into new business. He has served as a patent attorney in industry and in private practice and for several years worked as a business development manager for new ventures and technologies in North America, Europe, Africa, the Indian subcontinent and the Far East.

Peter gained his professional qualifications while working with specialist chemicals company Laporte Industries Limited. The company's unique developments in producing hydrogen peroxide and titanium dioxide pigments drew him into a series of lucrative licensing deals that funded acquisitions of other chemicals producers. At the same time the intense competition between Laporte and rival suppliers brought him into several patent actions in the courts, up to and including the Court of Appeal.

The business management period came while working with industrial gases giant BOC (now merged with Linde AG). In its New Venture Secretariat he helped launch and run businesses in cutting and welding equipment, fish farming, computerised brewing systems and in small scale gas separation plant. His best known sales negotiation was the first *Thames Bubbler* – a barge-mounted oxygen plant to protect the River Thames against pollution overflows.

Then came several years of bringing to BOC's subsidiary and licensee companies in over 50 countries the technology proven in its UK and US operations. This included placements and training courses for overseas personnel and frequent trips to the receiving countries to ensure the smooth launch and running of the new procedures. His successful course on *Skills of Negotiation* ran for several years at BOC's UK staff college. He was deeply involved in mergers and acquisitions, being part of the first teams to establish BOC's core gas business in Japan, Korea, and Taiwan.

Despite the successes in the Far East he became frustrated by the huge political difficulties in getting Western technology into much of the developing world and returned to a professional IP role in a London firm of Attorneys deeply involved in major court actions, including the 10-year long patent battles between British Coal and his clients, the Belgian glassmakers Glaverbel. This did not stop the globetrotting as in addition to trips to Glaverbel and other major clients in Belgium he acquired major clients in Germany, Japan, and the USA, and made regular visits to them.

He is a frequent adviser to new inventors in presentations and consultations to business groups, in one-to-one meetings at the UK-IPO and British Library, and latterly as an IP consultant to Trevor Baylis Brands plc. His sound common sense in guiding his clients led BOC patents manager George Renton to dub him 'the wisest man I've ever met'.

Philip Robinson was offered £50 for his final-year design prototype at the annual Bournemouth University Design Festival in 2001. Instead of taking the money he naively took the decision to commercialise his idea, thinking that within four weeks he would be enjoying the high life! Four years later and with a wealth of first-hand experience behind him, his patented *ZorinPump* successfully made it to market and is now on sale in over 30 countries.

The Biologic ZorinPump is an ingenious combination of a seat post and powerful floor pump. The pump mechanism is hidden inside the seat post so the pump is always at hand and hidden from the elements and thieves. It can inflate a tyre in as few as a dozen strokes. Ergonomically, the ZorinPump is just like a floor pump: you stand while you pump and use the saddle as a handle. Find out more at www.zorinpump.com.

In 2005, after a considerable period of due diligence and negotiation, Philip signed an exclusive worldwide production and distribution agreement with the world's largest manufacturer of folding bikes, Dahon Incorporated. His ZorinPump is now built into Dahon bikes as standard equipment and available as a retrofit accessory to fit other bicycles.

The commercial success of the product has been complimented by significant media interest and public exposure. It was featured in the Thomas Heatherwick & Sir Terence Conran Exhibition of Innovation at the Design Museum, London and also received a Department of Trade & Industry 'Smart' innovation award. Philip's personal achievements include becoming a finalist in the Daily Mail / Lloyds TSB 2006 Enterprising Young Brits competition and being nominated for the 2007 Morgan Stanley 'Great Britons' awards.

Amongst his other business interests, Philip is now a champion for product design students to encourage the successful commercialisation of good ideas. He speaks publicly about his experiences in product design and development, finance, licensing and commercial negotiation, intellectual property, business plan creation, and invention assessment. He has been a judge and guest speaker for the UK Intellectual Property Office's student 'ThinkKit' initiative and supports the National Council for Graduate Entrepreneurship Flying Start programme.

Pamela Riddle Bird is a recognised commercialisation expert who's counselled thousands of inventors and entrepreneurs over two decades.

After directing one of the largest publicly funded innovation centers in the United States, Dr Bird founded and serves as CEO of Innovative Product Technologies, Inc., a product- and technology-based market commercialization corporation. Dr Bird works with independent inventors, serving as a liaison between inventors and inventor organisations, venture capital organisations and other investors, manufacturers, entrepreneurial networks, and research park facilities.

Dr Bird is the author of more than 70 publications and has been quoted and featured in numerous newspapers throughout the United States including *The New York Times, Barron's – The Dow Jones Business and Financial Weekly, Forbes Magazine,* and the *Miami Herald.* She's a featured speaker in a video titled *Inventing, Patenting and Profiting: How to Make a Fortune on a Small Budget by Inventing.* She has been a guest on various US TV shows, including *Golden Lifestyles,* and she appeared in and served as a consultant to ABC's *20/20.* She's also taught classes on product commercialisation and technology transfer at various universities throughout the nation.

Dr Bird worked with MBNA (one of the largest credit card companies in the world) to start the first credit card for innovators, patent attorneys, and patent agents. All proceeds Dr Bird receives from this card are donated to the Inventors Educational Foundation. Dr Bird founded this nonprofit foundation that assists innovators and entrepreneurs of all ages and walks of life with educational and commercialisation needs.

Among her many memberships and organisational affiliations, Dr Bird served under three governors as commissioner for the Governor's Commission on Women and an adviser for the Adult Community Education Board and also the Regional Coordination Council. She's been the recipient of various recognition awards, including the Outstanding Community Service award, and has received letters of appreciation for community involvement in labour employment issues, child abuse prevention, crime prevention, and education needs.

Dr Bird is a pilot who also enjoys horse riding, snow skiing, hiking, fly fishing, and working with youth science fairs.

Authors' Acknowledgements

Peter Jackson and Philip Robinson: Our first and greatest acknowledgement must go to Dr Pamela Riddle Bird, the author of the original *Inventing For Dummies* published in the United States in 2004. With the help of a number of gifted innovators and individuals, to whom we also give our grateful thanks, she created a classic work for inventors and all those concerned with taking new ideas from concept to successful marketing. Our work in producing this British edition was made so much easier by her efforts in presenting a complex subject in a clear, logical, and very readable way. This version retains most of the structure and much of the wording of her original, while of necessity making substantial changes to take account of the differences between the relevant US and European intellectual property laws and of the different local institutions that are available to help inventors on this side of the Atlantic.

Our thanks go also to the people who gave us the opportunity to produce this edition: to Commissioning Editor Samantha Spickernell for inspiring us to take it on, to Assistant Editor Wejdan Ismail for her enthusiastic encouragement and to Development Editor Simon Bell for guiding us through the whole process of putting it together. And thanks to our colleagues John Grant and Charlie Ashworth for their help and soothing words as the various submission deadlines were bearing in upon us, and indeed upon them as they worked on their parallel work *Patents, Registered Designs, Trade Marks & Copyright For Dummies*.

We are delighted to have the opportunity to thank those who helped us start on the innovation road: to the now retired Bob Pitkin at Laporte and Michael Dowler at Abel & Imray, and also to Mark Taylor of Lloyds TSB, Geoff Bell of SaxonKing, Phil and Peter of DesTech (UK) Limited, and Oliver Blackwell, Steve Green and James Coulter for their encouragement and support.

Thanks go too to our families for their support and their patience in awaiting household and family activities that were put on hold over the months of putting this work together.

Finally we are deeply grateful to Miles Rees and Lawrence Smith-Higgins and their teams at the UK Intellectual Property Office both for letting us draw from their excellent range of IP guides and for undertaking the daunting task of reviewing the fine technical detail of the entire work. We know that they share with us the aims of increasing awareness of intellectual property and how to secure it, develop it and benefit from it. We hope that by providing in a single reference work a guide to the whole business of inventing from concept to conclusion we have made a useful addition to the library of innovation.

Pamela Riddle Bird: When a work of this magnitude is released, you can be assured that a number of individuals were heavily involved; therefore, I am indebted and thankful to many. First, I want to thank the innovators who make this book possible and who have changed the world.

I want to thank my husband. Forrest is not only my husband; he is my best friend and soul mate. He's a man who's dedicated his life to saving other people's lives and is a man of honur who served in three wars and still continues his mission of inventing for mankind.

Undoubtedly, I will leave out some people who were important to the project, and for that I apologise. My official thanks must start with my official Dummifier *par excellence*, Ms Kathleen Dobie. Kathleen is an incredible person and a delight to work with. She has insight and is an exceptional editor. I also want to thank my Project Editor, Ms Chrissy Guthrie, for her drive, energy, and talents in making this book happen, and Copy Editor Jennifer Bingham for checking and verifying everything to make sure it's right. Ms Kathy Cox, Acquisitions Editor, believed in me, my talents, drive, and work and was the champion for this project and instrumental in getting it off the ground.

I would like to especially thank Gerald G. Udell, PhD, and Mr Donald G. Kelly. In addition, I received much of the legal advice, not only for this book but over two decades of working with inventors, from the following intellectual property attorneys: Mr James Beusse and Ms Christine McLeod of Beusse, Brownlee, Wolter, Mora & Maire, PA in Orlando, FL; Mr Robert Downey of Robert M. Downey, P.A. in Boca Raton, FL ; Mr William Hobby, III in Winter Park, FL; Mr Robert Kain, Jr of Fleit, Kain, Gibbons, Gotman, Bongini & Bianco in Ft. Lauderdale, FL; Mr John Kirk, Jr of Jenkens and Gilchrist PC in Houston, TX; Mr Peter Loffler in Tallahassee, FL; Ms Jennie Malloy and Mr John Malloy of Malloy and Malloy, PA in Miami, FL; Mr John Oltman of Oltman, Flynn and Kubler in Fort Lauderdale, FL; Mr Thomas Saitta of Rogers Towers Bailey Jones and Gay, PA in Jacksonville, FL; Mr David Saliwanchik and Mr Jeffrey Lloyd of Saliwanchik, Lloyd and Saliwanchik in Gainesville, FL; Mr Jesus Sanchelima in Miami, FL; and Mr Brian Steinberger of Law Offices Of Brian S. Steinberger, PA in Cocoa, FL. I would also like to especially thank Mr Craig Dahlin, CEO of MarketreaderPro.com, and the following: Mr Eugene Andrews Grinstead IV, Ms Kristine Homant, Lowell Salter, Mr Ted Schaewecker, Ms Joanne Hayes-Rines, Mr Edward Miller, and Mr Robert Loughler.

Finally, I am ever so grateful to the board members for my company: Dr J. Robert Cade, MD; Mr Lloyd Bell; Dr Forrest M. Bird; Mr Philip D. Bart; Mr Edward Shadd; Mr John Weber; Mr Harris Rosen; Mr Patrick Perry; the late Mr Edward Lowe; and the late Dr Jay Morton. These innovators and entrepreneurs believed in me when I started my own business dedicated toward working with those who want to change the world. They've directed me through the thick and thin. They, too, believe in the undying spirit of the independent inventor.

Publisher's Acknowledgements

We're proud of this book; please send us your comments through our Dummies online registration form located at www.dummies.com/register/.

Some of the people who helped bring this book to market include the following:

Acquisitions, Editorial, and Media Development

Project Editor: Simon Bell

Content Editor: Nicole Burnett

Commissioning Editor: Samantha Spickernell

Developer: Kate O'Leary

Copy Editor: Charlie Wilson

Technical Editors: United Kingdom Intellectual Property Office

Publisher: Jason Dunne

Executive Project Editor: Daniel Mersey

Cover Photos: © Paul Block/Alamy

Cartoons: Ed McLachlan

Composition Services

Project Coordinator: Erin Smith

Layout and Graphics: Reuben W. Davis, Melissa K. Jester, Stephanie D. Jumper, Laura Pence, Christine Williams

Indexer: Claudia Bourbeau

Brand Reviewer: Jennifer Bingham

Contents at a Glance

Table of Contents

Foreword

Many of us have an inventive idea within us but lack the information needed to protect it and to gain its deserved rewards. All too often inventors fall prey to cowboys who give them misinformation and turn them over like a proverbial turkey.

Our society depends on creativity and must acknowledge, help and reward the people who provide it. They are amazing people and they have the ability to change all our lives, both commercially and socially. UK plc is good at inventing but appalling when it comes to looking after its inventors and inventions. Frustratingly in this country we spend more money on art than on invention. As they say, art is pleasure, invention is treasure, and this nation has got to recognise that. If it can spend a fortune on dead sheep and formaldehyde, then it can spend a bit more of that money on inventors whose ideas represent our future.

Unfortunately lack of respect and reward for inventors goes back a long way. John Hargreaves' Spinning Jenny was pirated and he lost both his machine and his house. Sir Christopher Cockerill, who invented the hovercraft, couldn't afford to take his family on one. Sir Frank Whittle is my favourite inventor. He invented the first jet engine in 1937. Unfortunately, no one listened to him so the Germans ended up using jets before the UK during World War II. If people had listened to him earlier the war might have finished sooner.

Sadly too, most people can't name any female inventors, bar Marie Curie and a few others. Yet the actress Hedy Lamarr invented a communications system that in time led to the mobile phone. Stephanie Kwolek invented *Kevlar*® but her name is almost unknown. There are not many women who are given the opportunity to reach their full inventive potential.

Such neglect is still more likely than not to happen. Many British inventions go overseas. If the country is not going to lose entitlement to its inventions, it's got to stand by you the little guy and your idea. First it must help you to bring out that idea within you. Most inventors are born with the need to invent. Such drive should be encouraged and not dismissed, as it often is; after all achievement is more important than qualifications. The key to success is to risk thinking unconventional thoughts and to follow your heart. It's better to entertain an idea than live with the regret the rest of one's life. It's not an option to die without leaving a legacy to others - after all there are no pockets in a shroud.

My attitude is that there's an invention in all of us. If you can solve a problem, you are on the way to being an inventor. Every now and then there's something that's unique, so you've got to recognise when that comes along and do something about it.

Society must guide you so your invention doesn't get stolen. It must help you record the invention. No-one will pay you for just an idea, but they may pay you for that official piece of paper which shows you have an idea with some rights attached to it. That official piece of paper also gives you the chance of a day in court to defend yourself against the sharks. But remember that to be an inventor you need an ego the size of a truck and have to be prepared for a rough ride. Invention is frequently one per cent inspiration and 99 per cent litigation.

Inventors clubs can and do help. They encourage and give confidence. So does the UK Patent Office. The rapport between the little inventor and their Private Applicant Unit is magic.

When it comes to bringing the products to market we in the UK get careless. We've got to make it easier for inventors to get to make it to the market. It's not just building the prototype, it's showing off yourself and your invention. You have much to do to fund the development. You must raise the money from somewhere but you need help to find it because you don't know who to trust. We must make sure that when the money rolls in you aren't rolled out, so you can make a living from the invention.

So I am very pleased to add my name to this book. It is a great step into meeting the need to educate you the inventors on the arduous road from idea to riches. It aims to take you all the way from that gleam in the eye idea to a successful product bringing you respect, rewards and recognition.

I have worked with both the editors of this edition in *Trevor Baylis Brands plc*, a company I formed in September 2003 with a group of experienced business professionals to help inventors learn more about their inventions, how to go about protecting them and seek routes-to-market for the commercially viable ideas. Peter Jackson is a patent attorney with huge experience and a long history of successfully spreading the word on the needs and benefits of intellectual property. He and I have shared a platform on a number of occasions to give face-to-face help to business audiences. Philip Robinson brings the younger man's approach with the impressive background of having taken his product idea from its University days' origins all the way through that challenging path to success in the worldwide market place.

Read, enjoy and learn from their writings, and keep these by you to refresh and encourage you through all your innovation efforts.

May all your dreams be patentable.

Trevor Baylis, OBE

Introduction

*I*t's been said that everybody has a novel inside them: A life story, a tale of broken and mended relationships, a gripping yarn of adventure in remote places. It's probably equally true that everyone has an invention to offer. The British have certainly shown themselves to be an inventive bunch, widely recognised to have made more world-changing inventions than any other nation. But all too frequently we have let others draw the financial benefits from our ideas.

In the US original version of this book, author Pamela Riddle Bird suggested the primary reason for inventors' failure to bring ideas to market was not a lack of funding but rather that inventors are so blessed with creativity that they just can't get focused, not knowing which of their inventions to concentrate on. While agreeing with these sentiments we must add that inventors in the UK and the rest of Europe seem to have problems not shared by most inventors in the United States: They generally have a much more limited knowledge of patents and other forms of intellectual property and they have less confidence in taking the undoubtedly risky steps of taking their inventions out of the shed or laboratory and turning them into business opportunities.

We hope that this book will help you overcome any such problems, give you a thorough briefing on innovation and the laws and organisations that support it, help you focus on your best ideas, tell you how to protect them and, most important of all, encourage you to make the most of them.

About This Book

This book is designed to answer your questions about how to take an idea and turn in into a product. Put another way, it can help you turn your dreams into reality. To do that it puts together our collective experience of bringing inventive ideas to fruition.

We tell you how to check whether your idea is new and protectable. We fill you in on the steps you need to take to develop and bring your product to market. We point out potential funding sources and tell you how to get in touch with a vast array of people who can help advance your project. We also alert you to possible snags and help you avoid common pitfalls.

This help is offered in an easy-to-read, easy-to-access format. Each chapter stands alone, serving as an individual piece of the whole inventing-and-marketing pie. You can dip into any chapter or section that interests you, then skip on to the next topic, whatever and wherever it is. You may be interested in reading some chapters more than others; however, in the long run, you will need the information in all of them.

Throughout the book we explain concepts that may be new to you and give you information and advice in clear, straightforward language.

Conventions Used in This Book

This book uses a few conventions that you should be aware of:

- ✔ It uses *italics* either for emphasis or to highlight terms and concepts explained in case they are new to you.

- ✔ The stories in grey boxes are known as *sidebars*. Sidebars contain information you may find interesting or useful, but which you don't need to understand the topic at hand. You can choose to read them or not.

- ✔ Web sites and e-mail addresses appear in a different font `monofont` to help them stand out in the text.

Foolish Assumptions

We assume that you're reading this book because you have an idea and want to know what to do next. You want to find out whether your idea is marketable and how to get it to market and make a profit from it.

Whether you have a prototype or a patent yet doesn't matter. Maybe you have both and want to know what to do next. You want to move forward and do something with your idea. You not only want to see people buy it but you want to make money as well.

Rest assured that you've come to the right place.

How This Book Is Organised

Inventing For Dummies is organised into seven parts. The chapters within each part cover specific topic areas in detail.

Part I: Assessing Your Idea

This part shows you how to assess your idea step by step. It gives you the security of knowing what to discuss, with whom, and how to keep those conversations confidential.

These chapters review the many types of protection that may be available to you: Patents and other kinds of intellectual property. They tell you how to discover if your idea is the ground-breaking advance you'd like it to be – how to search to find out if it's already out there.

Part II: Protecting Your Idea

The chapters in this part tell you how to use patents and other kinds of intellectual property rights to protect your inventions, words, designs, pictures, processes, brand names, and trade secrets. They show you how to apply for a patent, and how to take care of your patent after you have it.

But many innovations aren't patentable products. So this part also guides you to other IP rights: some that come your way automatically and some you have to apply for. It describes automatic design rights and other forms of copyright, and the possibility of officially registering your designs and trade marks, and how to take care of them too.

Part III: Developing Your Idea

The next time you pick up a product that's new to you, take a step back and think about how that product got to the shops. The inventor had to first visualise the idea, try it out – maybe building a working model to see if it would work and if it looked right, test whether people would buy it, figure out if it could be made for the price customers would pay, get it produced, and have it delivered to the retailer and end user.

The chapters in this part of the book take you through these development steps, both for new products and for improved processes.

Part IV: Preparing to Enter the Market

Very few people understand the process involved in bringing a new product to market. After reading the chapters in this part, though, you can be one of them.

The invention process is quite different from the commercialisation phase. Most inventors have either to license their intellectual property rights to a third party or become entrepreneurs themselves.

Becoming an entrepreneur requires a whole new knowledge base, skill set, and often a change in lifestyle. These chapters help you understand the risks and rewards of starting a business to produce or sell your idea. They guide you on creating your business plan, finding funds, choosing associates you can trust, and getting ready to launch into the market.

Part V: Developing Your Market

Getting a toehold in the market is just a start to getting worthwhile returns. The business needs to grow from that start. You need to promote your business, advertise it, maybe license or franchise it. These chapters show you how to expand the business and how to negotiate the licensing process from contact to contract.

Part VI: The Part of Tens

As a reassuring postscript to the many issues you face in turning your idea into a commercial success, these short and sweet chapters list helpful organisations for inventors and include brief details of successful inventors and inventions that had an impact in the past and continue to do so.

Part VII: Appendixes

This part provides forms and resources that you can use to protect your idea and find the help you need to make it a success.

Icons Used in This Book

The icons you sometimes see in the margins of the pages are there to draw your attention to the text they're next to. This book uses four different icons:

This icon directs your attention to information that can make your efforts easier or more effective. Pay attention.

Make note of the information next to this icon; the stuff is important and you'll need it.

This means watch out! Pay attention so that you don't make a mistake that can hurt you or your wallet. It marks things to avoid and common mistakes inventors make.

This icon highlights technical information. You can skip this information if you like, but just because it's technical doesn't mean you can't benefit from it.

Where to Go from Here

We recommend that to get an overview of the whole inventing process you first read Chapter 1. Beyond that, look through the book and find which part, chapter, or section you're most interested in. And carry on from there.

This book is written in such a way that you can turn to any section, in any chapter, in any order you like, and never feel that you've missed a trick. If you want to read about intellectual property issues, check out Parts I or II. If you want to know about development details, turn to Part III. If you're interested in the sales and marketing aspects, go to Part IV. Or, start at the end and read about and be inspired by ten individuals who changed the world.

This book is written for you, and by reading it, you're one step closer to making your dream into a reality and turning your idea into a marketable product or technology.

Now, you have the know-how in your hands. Take advantage of it and good luck!

Part I
Assessing Your Idea

'So this is your invention that helps the
most hopeless adult with today's modern
IT technology.'

In this part . . .

The first step on the road to a successful invention is to decide how good – and how original – it is. This part takes you through that process step by step. It also shows you what to discuss with whom, and how to keep those discussions confidential.

These chapters review the many types of protection that may be available to you, and tell you how to find out whether your idea is the ground-breaker you'd like it to be.

Chapter 1

The Innovation Process

*Y*ou can dream of waking up one morning with a great idea, making a phone call, receiving a cheque by noon, and reaching financial success, all in time to watch the sunset through your lovely rose-coloured spectacles. It may be possible. It's possible to win the lottery, too. It's possible to earn both an Olympic gold medal and an Oscar.

Dream on! Being possible doesn't make it likely, just as getting a patent doesn't make your invention commercially viable. Try to accept right now that although making money from your idea may be possible – potentially a great deal of money – achieving that happy result may be a long, gruelling, and expensive process. Many people never see their idea become a reality. Few see a good return on their financial investment, much less of their time. Only a small percentage of patent holders ever make enough money to recoup even the cost of getting their patents.

But don't give up on the dream yet. You're satisfied that you've got a great idea for a new product. You think that the public will want to buy it. Give plenty of thought to your journey of taking it from being just a gleam in your eye to becoming a finished and packaged product flying off the supermarket shelves. Set yourself a programme of the steps you face in that journey, visualising the milestones you have to pass. Be persistent. Prepare to recognise little successes along the way as victories. So, as you reach the milestones that you've set for yourself, be happy and proud of these victories. Each one is getting you closer to that pot of gold at the end of the rainbow.

This chapter gives you some practical advice on making, protecting, and improving your invention. It also introduces the product life cycle concept to give you a better idea of the steps you need to take, including a few basic pointers on finding the best route to market, before later chapters go into greater detail.

Deciding Where to Go with Your Idea

So you have your idea, now what? Where do you go? What steps do you need to take to bring it to fruition?

Decide as early as possible what your true objective is. In other words, what do you want from your invention? Do you want to be famous? Is just getting a patent what you really care about? Are you all about the money? Are you all about making the world a better place?

Your goal plays a large part in how you proceed with your idea. If your true desire is to have your name on a patent and commercialisation isn't important to you, your task is substantially different than if you're expecting your invention to bring you financial independence.

But if you like the thoughts of commercialising your invention to the point of bringing in great cash returns, you're the type of person this book is aimed at. You play a great role in society. The wealth created by people like you doesn't just give you a better-paid lifestyle. It gives employment to designers, manufacturers, shippers, distributors, retailers, and advertisers, and to the shops, coffee bars, pubs, and restaurants where they spend their money after work. And you're in good company. When Thomas Edison, one of the greatest inventors of all time, couldn't find a buyer for his first patented invention – an electric vote counter – he formulated a lifelong policy: 'Anything that won't sell, I don't want to invent!'

Carefully consider where you are today and where you want to end up. It may be helpful to take a sheet of paper and list the steps involved in achieving your final objective. Try writing where you are today at the top, skipping several lines, and entering your final goal. Now enter the milestones that you consider to be important. Some of the tasks you may list are

- Conducting a patent search (see Chapter 4)
- Submitting your idea for evaluation (see Chapter 12)
- Filing a patent application (see Chapter 5)
- Putting your idea into practice (see Chapter 10)
- Researching the market (see Chapter 18)
- Forming a company (see Chapter 16)
- Meeting with a potential licensee (see Chapter 21)
- Writing a business plan (see Chapter 14)

You don't have to fill in all the blanks if you aren't sure of all the steps; having a rough idea of some the steps is enough to get started (you may come up with additional ideas as you go through this book). But preparing even a preliminary list may make you realise that there's a lot more to the process than you had originally anticipated and that you may need help along the way.

You need to continually update the list. If you have a target date for success, then include that as well, and then date the milestones that must be met to achieve it.

Protecting Your Idea

One of your first objectives, regardless of your overall goal, is to make your idea yours. In order to do this, you have to ensure that your idea is protected. That's what Parts I and II of the book are about: officially protectable rights such as patents, registered designs, and registered trade marks, and other rights that come to you automatically, such as copyright, design copyright, and certain common law rights. Collectively these rights have become rather grandly known as *intellectual property*, IP for short. It's a term for them that was hardly known a few years ago but it's useful and becoming more commonplace, and for convenience is much used in this book.

If you're only interested in getting a patent or other IP rights, stick to the chapters in Parts I and II. The rest of the book, however, focuses on experiences and expertise in bringing inventions to market. If you want to make money from your idea, and we hope that you do, every chapter and every section should be of interest to you.

Safeguarding your intellectual pr0perty

Your idea – or at least your rights to your idea – can slip out of your grasp if you don't protect it well. Just bragging too specifically to the wrong person can be dangerous – people claim credit for ideas that aren't theirs all the time. So though you may be proud of your idea, be careful about how you talk about it and to whom. If you think about it, it isn't logical to share your brainstorm with the world without safeguarding it.

So give some early thoughts to seeking and using intellectual property protection – a patent, copyright, or other rights mentioned in this part. Taking precautions right from the start can help bolster your claims to protection and firmly establish you as the rights' owner.

Keeping good records

Keeping an inventor's diary, engineering notebook, or logbook can be a great help, not just to assist in defining and protecting your rights, but to document your invention's history and progress.

Detailed documentation can be a valuable resource in many stages of your invention's life cycle. Potential investors interested in funding your project can use your records to make decisions about your product's potential and about your own capabilities. Investors are often more interested in the person than in the product. If, for any reason, or no reason, the Inland Revenue audits you, your files can show and justify deductions you may have taken.

What if an employee claims that he or she is the actual inventor or maybe a co-inventor? Or your lab suffers a break-in, and whether the thief was hired by a competitor or is just an opportunist, you may lose your prototype and equipment. In cases like these, a well-kept invention log can help you pick up the pieces. But take care to keep it in a safe place and regularly make a back-up copy.

Systematically recording your progress can be a useful discipline in developing your ideas into a viable invention. Consider including

- ✔ The title of your invention
- ✔ The purpose of the invention – what you want to use it for
- ✔ A detailed description, including any unique features
- ✔ Sketches, drawings, or pictures
- ✔ Alternative uses and applications for your product – these may vary by industry
- ✔ The differences between your invention and similar products, if you know of any
- ✔ Advantages of your product compared to other products or close substitutes
- ✔ Names and contact details of the people you speak to about your product, including consultants, prototype builders manufacturers, packaging designers, potential licensees, and so on; and names of people who you know are aware of your product, including employees, friends and family, and current or former colleagues.

It helps to document what you and any colleagues do to further your invention, and to keep records of any correspondence, receipts, and bills related to the project. For example, you may have correspondence about manufacturing cost estimates, potential licensing agreements, hiring an engineer or prototype builder, and so on. The record of expenses also comes in handy at tax return time.

Cheering on inventors

Inventors are frequently made fun of and ridiculed. The general public tends to think of them as wild and crazy eccentrics, best avoided. That is, until one of their inventions sells. The wild-eyed madman suddenly becomes a hero, the salt of the earth, the kind of person who makes the country great.

The truth is that inventors have changed and continue to change the way we live. The vehicles we drive, the medicines we take, the clothes we wear, the computers we use – the list is long of everyday inventions we use with scarcely a thought to the people who invented them. Inventors contribute to society, often with no idea of how great an impact their creations can have on generations to come.

One inventor who was asked 'Why do you invent?' replied 'Well, I never could be God, but I love to create.' Deep down, we all like to help each other make life easier, and inventors do

make a difference. Successful inventors have the same characteristics as any other successful individual. They have a strong interest and belief in their goal and a persistent drive to reach it.

Although our base of technical knowledge has increased exponentially in modern times, revolutionary inventions don't have to be complicated. Although the paper clip and the safety pin seem fairly simple pieces of engineering, both continue to have a huge impact on everyday life.

Inventions, like inventors, come in all shapes and sizes. You don't have to have an advanced engineering degree to be a successful inventor; a major US study revealed that an inventor with a university PhD degree is only a third more likely to bring an invention successfully to market than an inventor with a high school diploma.

If your invention is very good, you almost certainly face having it copied. This is where your intellectual property rights come into play. In legal terms the people with copies are infringing your rights. In an ideal world your attorney's letter to the infringers advising them of your rights will make them back off. More often though, you or your legal advisers face exchanges or meetings with the infringers or their advisers to come to a negotiated withdrawal or perhaps a licensing agreement. At worst you end up in court to enforce your rights. Full background documentation can be a real help in getting a good defensible written definition of your invention into your patent specification and later in giving written supporting evidence to support your case in negotiations or in the courts.

Spinning Through the Product Life Cycle

You've come up with a practical solution to a common problem and you want to make money from it. Where do you start? Well, first of all, don't quit your day job and buy the Rolls Royce just yet. Your inventor's high hopes may soon come crashing down amid the frustrations and setbacks of developing your idea and getting it to market. You must consider many things before running to a patent attorney to file a patent application.

Having thought of their good idea, some inventors want to give it at once to somebody else to make money for them. But doing so is like handing someone your newborn child and asking them to feed, clothe, potty train, educate, and otherwise care for her until she's fully grown and through college. No one has more interest or investment in your invention than you have and to maximise the potential profits, you have to be involved in rearing it. Commercialising a product has been likened to having a baby – easy to conceive but hard to deliver.

A product, like a child, has a distinct life cycle. In today's competitive markets, you must not only have a great idea, but also know how to move it through the entire commercialisation cycle. And with the right kind of involvement you can enjoy every moment of it – well almost.

 Be careful not to get so committed to your idea that you overlook how much money its development is running away with, or how much time it's taking up when you ought to be out earning your daily bread. Inventors have been known to take out massive loans, remortgage the house, or otherwise get deeply into debt to pursue their goal. Some emerge at the far side with a huge market success and great financial returns (see the Mandy Haberman story in Chapter 25) but others don't get there and face financial ruin. Regularly review how the project is going and don't be afraid to cut your losses by dropping out if things start going badly.

The rest of this chapter looks at the typical steps in a product life cycle. The timing of the cycle varies according to the product. Once developed, simple gimmicky toys may only have a market for a year or so. Toy manufacturers put in huge efforts trying to make their product this year's must-have Christmas present. Others, like the steel can for baked beans, seem to go on forever, although even they sometimes get an update such as the ring-pull lid so you don't have to search for a tin opener. For many products however a typical lifetime is ten years from the gleam in the eye, through evaluation, development, patenting, manufacture, and marketing, to sales returns and then sales decline until hardly anybody wants to buy them any more.

Step 1: Idea generation

The dream of waking up with a great idea is not all that wide of the mark. Ideas come from unexpected sources and a number of these may well occur to you in the small hours. Even highly creative people can rarely produce a good idea to order. Telling someone to go into a room and invent something rarely produces anything worthwhile. Experiments asking groups of students to invent something have produced a high proportion of suggestions for a toothbrush with the toothpaste in the handle. (Okay, that may be useful but it was first thought of years ago and never really caught on.)

Many successful inventions have come about by accident, among them potato crisps, Velcro® fasteners, microwave ovens, Post-it® notes, vulcanised rubber, Viagra® tablets, and penicillin. So go round with your eyes open. Be on the lookout for an unexpected aspect of something familiar. Having an enquiring mind helps. Curiosity as to how something works can lead to thoughts of how to make it work better.

And if you favour a more systematic approach, think about problems that really bug people. Choose one and then look for and develop a new and improved solution to beat it. And don't be afraid to challenge the conventional wisdom.

Step 2: Idea evaluation

This book mostly assumes that you're beyond the idea generation stage. You've had your Eureka moment and thought of something new and useful that you're sure will go down a storm. Think first about how it's to be used. Think about it in terms of marketability. Can you turn it into a saleable item? If so, will people buy it? Is there something already on the market that's very similar? What other competition is there? Remember, ideas are cheap and products are expensive.

 The merit of your product isn't the only key to its success. So many factors can affect how your invention does in the marketplace. The most you can do is to cover all your bases as well as you can, and that's what this book aims to help you with.

You're receiving no income because you're not selling anything yet. You're testing feasibility, viability, and reality. You conduct a patent search to make sure that someone else doesn't already hold a patent on your idea. (Just because a product hasn't been marketed doesn't mean that there's no patent on it.) You do estimates of production costs, and conduct market research to find out whether consumers are interested and how much they'd pay.

Not so green . . .

A group of Japanese university research students charged with finding new superconducting materials were beaten by a rival group who produced an excellent superconducting compound that happened to be green in colour. The first team had previously arrived at the same compound. But their professor had told them that green materials wouldn't superconduct, so they didn't test it.

Step 3: Technical research and development

You make a prototype, or get one made. You look into constructional materials that are more robust/flexible/reliable/long-lasting than the old plastic sheeting you found in the corner of the garage. You look into simplifying the design to use less parts, or parts that are cheaper or more readily available. You try to make it smaller or lighter to do the same job. You consider its green credentials: how toxic/abrasive/recyclable are its components? What machinery is necessary to make it in commercial volumes? You improve its controls to make them more user-friendly. You try to make its appearance visually attractive.

At some point you reach a level of development at which you have sufficient details and awareness of the distinguishing features of the product to support preparing and filing your first patent application. See Chapter 5 for more on how to set about this. You don't want to file before you have a version with reasonably well proven inventive features, but on the other hand you don't want to wait too long before you file because there's quite likely to be somebody else working on similar lines and you don't want them to get a filing date before you do.

Practical engineers have great sympathy with Voltaire's observation that 'the perfect is the enemy of the good'. In other words don't always strive for perfection. Don't spend another £2,000 on developing an improved switch that may only save 5p in manufacturing costs when the basic switch works quite well enough.,

With the first patent application in place you may feel ready to approach an investor to put money into the forthcoming stages. Your project funds have by now been depleted by the development costs and new investment cash is likely to be extremely welcome.

It isn't necessarily the best product that ultimately succeeds in the market. A good product that has advantages in terms of clever advertising, exposure to the industry's movers and shakers, well-timed discounting, or smart point of sale presentation can sometimes succeed over better products. Think about VHS video recorders: they saw off Betamax even though the pundits thought the Betamax picture quality was far better.

Step 4: Production

With or without your investor you set up your factory and its assembly line, or more likely – and probably more sensibly – you find an established manufacturer to make the product for you. There may be, probably is, a need for preliminary production and testing stages to make sure that full-scale production reliably turns out first grade products in the required numbers.

Step 5: Introduction

Eventually, probably at least two years after you started working on your idea, your product enters the market and – with the help of your advertising and marketing – begins to sell and bring in returns to start paying off the bank loan, or at least the interest.

By now, and perhaps much earlier in the cycle, you may feel ready to back out and license the invention to a company to take on the project in return for payments to you, perhaps by way of a royalty. Or you may wish to back away just a bit and appoint or hand over to managers with expertise to move the product into the stages of market development and growth. Management teams come from various walks of life, with different fields of expertise and experiences. In fact, one of the most valuable things an investor brings to a company isn't necessarily funding, but the management expertise and experience to help push your product along.

Clearly, people with money to invest in your product don't have the funds because they're bad at business. No doubt they made mistakes, but that just means that they can advise you of errors to avoid. Try to build a solid and varied advisory team to help steer your invention down the most profitable path.

Step 6: Rapid growth

After introducing your product, you continue to develop the markets for it and, hopefully, witness rapid growth and soaring profits. As a product's market presence grows, it has to be competitive in order to survive. Marketing is civilised warfare! Your product will go through all sorts of growing pains, much like a teenager moving to adulthood. You have to be as creative as you were during the invention stage in order to take and hold a place in the market.

In the marketplace, your invention fights for the same pounds that can be spent on a family holiday, a mortgage, a new car, a child's education, and so on. Consumers only have so much money to spend and a variety of ways to spend it. Your product is in direct competition with other products and services. How consumers choose to spend their money must be analysed. Failure to do so and to understand consumer psychology can cause them to purchase similar products or close substitutes instead of your invention.

Although the product is going well, don't give up on investigating improvements to it. Much product marketing relies on this year's version having some or other improvement over last year's. New features can help keep you ahead of the competition, and they give the opportunity for further patent applications to boost and extend your legal rights.

Step 7: Maturity

Your product eventually moves into a mature stage in which sales peak, capturing as much of the potential market share as possible. This generally happens after about the fourth or fifth year of the product life cycle. Eventually, other new products with new technologies and improvements start competing in your market and your sales decline. This stage is one reason why companies constantly develop new and improved products.

Step 8: Abandonment

Finally, often between the seventh and tenth year, a product reaches an abandonment stage. It's no longer profitable to continue manufacturing, marketing and distribution. Your patent is still in effect but the patent alone isn't enough to keep your invention profitable. Go away on an ocean cruise, funded out of your handsome profits from Steps 6 and 7, and dream on deck about your next product idea.

Chapter 2

Keeping It Confidential

....

....

*P*rotecting your idea is an issue from day one. Disclosing their invention before protecting it is a huge trap into which many inventors fall before they even embark on the path to riches. That's why this cautionary chapter on confidentiality is so near the beginning of the book.

This chapter covers the hows and whys of ensuring everyone respects your confidentiality – from your work colleagues, suppliers, and designers to the company thinking about licensing your product, and even your family. And don't forget that the person who can do most damage by talking about the invention too freely, and too soon, is you.

Spilling the Beans about the Basics

The UK, the rest of Europe, and most other countries have firm rules, if often seeming a bit unfair. If you reveal your invention to anyone before applying for a patent, you make a public disclosure. And that premature disclosure is enough to stop you getting a valid patent. Without patent protection you expose your great new idea to exploitation by competitors who take it up for themselves and don't pay you a bean. At the very least you may find yourself sharing a market that could have been all your own. At worst you watch in frustration as a fast-moving competitor gets way ahead of you in the market-place and cuts you out altogether.

Governments, represented by their intellectual property office (often known as the *patent office*), grant you a patent as a reward for you telling them, and thus in due course the public, all about your invention. The reward is your right – for up to 20 years – to stop anyone else making, using, or selling the invention in the country that grants the patent. Governments grant this right because the system draws out a stream of ideas that become freely available to the public after the inventors have received the benefits of their protected rights. But if you reveal the invention before applying for a patent, no need exists for the Government to give you the reward.

Ideally, you file a patent application on your invention before you give its details to anyone other than your patent attorney. The Government office (in the UK the Intellectual Property Office, UK-IPO) acknowledges the filing by issuing you a written receipt that records the date and the official filing number. So in any later argument or dispute, you can confirm that you owned the invention at the date of filing shown on your officially recorded receipt.

Often, however, you don't have all the information you need to prepare a strong patent specification without first talking to specialists such as designers, materials suppliers, and structural engineers, or getting the reaction of retailers to its market prospects.

Fortunately – as in most legal matters – notable loopholes exist that help you avoid or escape from the trap of premature disclosure. This chapter deals with how to make a disclosure under such conditions of confidentiality that it does not count as disclosure at all. Also you may be able to rely on other kinds of intellectual property (IP): designs, trade marks, copyright, know-how, and trade secrets. Chapter 3 has more to say on the whole range of IP rights and the Part II chapters go into the details of how you can acquire and defend them.

The following options may help if you have shot yourself in the foot as regards getting patent protection:

- ✔ Apply for UK or European design registration – for the shape or ornamentation of the product – within a 12-month 'grace period' from the date of disclosure.

- ✔ Apply to register your trade mark at any time. Disclosing your trade mark before filing the application does not invalidate the registration.

- ✔ Rely on the copyright that arises automatically at the date of creation of the work, for example the copyright in the drawings from which your product is made.

- ✔ Keep confidential the know-how and trade secrets essential for making the product in a cost-effective way.

In the United States a 12-month grace period even exists for patents as such, giving a welcome opportunity to work on the fledging invention and publicise it before filing the US patent application. Some commentators imply that the UK offers a similar grace period for patent applications. It does not!

Although a pre-filing disclosure within the grace period does not invalidate a resulting US patent, it probably would invalidate an equivalent patent in counties such as the UK. Moreover, after disclosing your invention (and therefore ruling out UK patent protection), if you as a UK inventor decide to apply for a US patent instead you must obtain written consent from the UK Intellectual Property Office before arranging the US filing.

Confidentiality Agreements

A *confidentiality agreement*, sometimes referred to as a non-disclosure agreement, is one of the most common tools you can use to protect your idea or invention. This is a binding contract between you and the person or company that signs it, stating that the other party won't disclose any details of the idea or invention without your express permission.

When you discuss your invention with a potential investor, licensee, manufacturer, or customer, a nagging concern is whether they may try to copy your idea. Getting them to sign a confidentiality agreement can help ease your mind and protect your rights.

The confidentiality agreement puts the other party on notice that you believe that you have an idea or product worth protecting. Ensure that your agreement covers all the information you own regarding your idea, concept, product, invention, process, or service, including the following:

- ✔ Any trade secrets, know-how, internal workings, designs, drawings, and so on that you may divulge.
- ✔ Everything that is not in the *public domain*, in other words, all information not already disclosed or readily available to the public.

A typical example of a confidentiality agreement is given in Appendix A. Another good example appears in the UK Intellectual Property Office booklet 'Confidentiality and Confidential Disclosure Agreements (CDA)', also available as a pdf file on their Web site (use www.ipo.gov.uk/cda.pdf).

Often a simple confidentiality agreement along the lines of the Appendix A or UK-IPO examples is all that you need but if you're uncertain about what to include or how to word it, then seek professional help. The section 'Hiring an IP Professional' in Chapter 3 provides guidance on professional IP helpers, what they can offer, and where to find them.

A good possibility exists that the company you're approaching is working on a similar product. If so, it should let you know that upfront. Revealing a competing product or idea is the best way to avoid future disputes.

You don't always need a confidentiality agreement if you've already filed a patent application to protect your invention. However, you may want to ask for one anyway, just as an extra precaution, and especially if you plan to reveal any trade secrets that aren't described in the patent application.

After you and the third party sign the confidentiality agreement, you can on principle tell them as much as you choose. But be wary if you get the feeling that the third party is not fully respecting the terms of the agreement. In this event use your good judgement and the knowledge of your professional IP helpers (see Chapter 3) to decide whether to end your dealings with that third party and approach another.

Make sure that the proper people on both sides sign any agreement, confidentiality or otherwise. If an unauthorised person signs an agreement on behalf of a company, the agreement may be invalid. Find out who the company's authorised representative is, and for your part make sure that your secretary or spouse doesn't sign the agreement on your behalf by mistake.

Sharing With the People in Your Life

Do ask anyone you must talk to, even a friend or acquaintance, to sign a confidentiality agreement. The odds are against anyone close to you revealing the specifics of your invention, but the odds are quite high that if your invention is successful you'll have to defend your rights to it. In a legal dispute, the opposing party may ask for the names of people you talked to about your invention who didn't sign an agreement.

Securing dinner-table talk

Asking your spouse and your family and friends to sign confidentiality agreements may feel odd. If you need help in making them understand that you're not asking them to sign because you don't trust them, just show them this chapter. At the very least, impress upon them the confidential nature of what you're doing and tell them as little about its details as possible.

Take care that you don't spill the beans when talking to others. Waxing lyrical about your latest inventive triumph is all too easy at home or in a pleasant restaurant. Don't do it! Keep the circle of people who know your invention's details as small as possible. If someone wants to invalidate your patent or registered design, all they have to do is prove that you told someone its details other than under a confidentiality agreement, and outside of any grace period, before you applied for the intellectual property protection.

Signing on employees

In order to develop your idea or invention thoroughly and successfully you may have no choice but to consult with experts in specific areas: a design engineer, a prototype builder, a tooling expert, a packager, and a bookkeeper. Each of these consultants is privy to your sensitive information. Some of them may work, now or in the future, with your competitors. They may well start their own business with your inside information. They have the means and contacts to pass on your information, so asking each of them to sign a confidentiality agreement is essential.

Make sure that you also get agreements with any subcontractors that your consultants use. The consultants themselves may have confidentiality agreements with their subcontractors and those agreements help to protect you, but don't assume those agreements are in place.

Zipping your lips at work

Especially if you're a professional or semi-professional inventor, or researcher, you need to be aware of the pitfalls of revealing the results of your work. Time and time again innovators lose the right to patent their innovations because they write about them for a trade journal, or share their findings at a seminar, before filing their patent application.

Get confidentiality agreements from any college and university professors and researchers you employ as consultants. Many of them are prone to the pressures of the 'publish or perish' mentality of academia and revel in discussing their research with a scientific journal editor, overlooking the risk that poses to intellectual property protection.

Going Without a Confidentiality Agreement

Sometimes you may face a problem in getting a third party to sign a confidentiality agreement. You may encounter an investor who simply doesn't want to commit or a potential employee who doesn't want to be restricted. Signing a confidentiality agreement may place a company in a tough legal situation: if it ever legitimately and independently come out with a similar product, it's vulnerable to your suing it for copying your idea.

Running into Someone Who Won't Sign

If the company you want to work with won't sign your confidentiality agreement you have two choices.

✔ Not disclosing what you have, and thereby restricting the possibility of business with that company.

✔ Trying to work out a confidentiality agreement that gives sufficient protection for you and for that company.

Before you make your decision, consider whether that company is likely to commercialise your invention and whether you can make a business agreement with it sooner or later. The company's reputation is another item to balance. A well-respected firm doesn't earn a good reputation by conducting dishonest business dealings.

Although a certain level of risk always exists, you can typically resolve confidentiality agreement issues if you approach them in a businesslike manner. If not, and you still want to work with the company, you may decide that it's best to wait until you've filed a patent application.

Talking to The Big Boys

Large corporations often resist signing confidentiality agreements. Many companies and licensing organisations have a hard rule never to sign a confidentiality agreement because in the past someone sued them, based on a confidentiality agreement relating to that someone's product, for a similar product that the company did not know was in development. The representative of a company with multiple research and development labs may in all honesty not know if one of its many labs is developing something similar.

For organisations that are in principle prepared to sign, their own lawyers often have to review the agreement before signing. Other firms have their own standard agreement that they ask you to sign before reviewing your offer.

Many large companies have a set procedure for dealing with unsolicited offers of new product and process improvements. Unless it is clear that you have a patent application with a specification that covers the material offered, and that the company is allowed to read the specification, it returns the offer unopened and unread. From the viewpoint of the large company, doing so removes any possibility that you may return to it in future to argue that its latest product or process is a copy of your proposal.

A confidentiality agreement can be a valuable tool for getting underway with third parties on design, materials, configuration, packaging, and all the other things that go into making a new product or process. But if your timing and resources permit, the safest route is to apply for your intellectual property rights before approaching any third party. (See Chapter 3 and the chapters of Part II for more on IP rights and how to apply for them). But if, for whatever reason, you can't apply for IP rights, a well-drafted confidentiality agreement is a good second best. And do be ever mindful of keeping mum about your bright new ideas, even at home.

Chapter 3

Intellectual Property Basics

*W*hat exactly is intellectual property (IP), and what can it do for you and your invention? Who issues patents and how long do they last? How do you register a trade mark or an attractive design? If you're asking yourself these questions (and maybe others too), you've come to the right place.

This chapter gives you the lowdown on the legal rights that are available to you, and why legal rights are so crucial to making money from your invention. Next, because your first steps into gaining rights for your intellectual property may go no further than a patent application in your home country, we tell you how to extend your protection into other forms of IP and into other countries. This chapter also indicates how IP professionals can help take you through the process of securing those rights.

Getting to Know Intellectual Property

Don't be put off by the term 'intellectual property'. This is just a convenient umbrella to cover the wide variety of property that inventors, authors, artists, and musicians create: in other words, the creations of the mind or intellect.

Like any other kind of property, IP gives its owners certain legal rights. Some of the rights arise automatically and others you have to apply for. They come in four main areas: patents, designs, trade marks, and copyright. We discuss each kind of right later in this chapter in a little more detail, and in Part II of the book at least one full chapter is devoted to each area. Here are broad definitions for each of the main areas.

- ✔ **Patent.** A right protecting the technical, structural, or functional features of your invention, covering how your invention works, how it's made, or what it's made of.
- ✔ **Design right.** A right protecting the appearance of your product, as a registered design or as an unregistered design (see the sections later in this chapter that explain the differences between the two).
- ✔ **Registered trade mark.** A right protecting your brand name or other product identifier.
- ✔ **Copyright.** An automatic right protecting you against anyone copying your artistic, literary, musical, or other creative work (we mostly consider copyright in this book with regard to your technical sketches or drawings and your computer software).

This section describes each of the above forms of protection in turn.

Patents

A patent is a legal right that a government office grants you – as the inventor of a new and improved process, machine, or other product – for you alone to have the right, for as long as the patent is valid, to make, use, or sell that process, machine, or other product.

Who grants patents?

IP offices set up by governments grant patents. The United Kingdom Intellectual Property Office (UK-IPO) grants patents in the UK. The Irish Patents Office covers Ireland. (Appendix B gives contact details for UK-IPO and the Irish Patents Office as well as other IP offices).

Knowing what a patent does and does not give you

On the positive side, a patent gives you the right, in the country that grants your patent, to stop others making, using, offering for sale, selling, or importing your invention for a fixed period of time. And a patent gives you the right to take court action to sue others for *infringing* the patent, which means utilising your patented invention without your permission.

But on the negative side, a patent does not guarantee that your invention will make money. Nor does it show that the product or process is necessarily superior to what is currently available, although it does indicate that the product or process represents an inventive step forward.

A patent doesn't give you the right to put the product or process of the invention on the market. This can be a difficult point to accept. The problem may be that your invention, although useful and patentable, falls within the scope of somebody else's existing patent. You can't offer your invention without infringing his patent and he can't offer your improved version without infringing yours. The solution is often to reach a cross-licensing deal: you license him to work within your patent and he licenses you to work within his. See Chapter 21 for more on patent licensing.

The granted patent

The granted patent includes a specification with a detailed description of the invention, technical drawings if appropriate, and most importantly a formal set of claims that define the differences and improvements that have made it patentable over what has gone before. The claims define your rights: what other parties can't do without your permission.

The IP office issues your patent for a term of 20 years from the date on which you apply for it, subject to your keeping it in force by paying renewal fees.

The United States Patent & Trademark Office grants US patents not only for new and improved processes, machines and so on (which it calls 'utility patents'), but also for new designs and for new plant varieties (called respectively design patents and plant patents). In most other countries, and as discussed below, protection for designs comes in the form of registered designs or automatic design rights; for plants, protection comes as 'plant breeders rights' or 'plant varieties rights'.

Enforcing your patent

After you receive a patent, *you* must enforce it. The IP offices won't enforce your patent rights for you – not the police, local, or national government. Chapter 6 contains more on enforcing your patent rights.

After your patent expires, your invention is part of the *public domain*, which means anyone can make, use, or sell the invention without your permission and without paying you anything.

Knowing what you can patent

In order for an invention to be patentable it must fall under the laws on patents; not just the Patents Acts but also related official rules and directions (see Chapter 5 for more on these). Subject to these laws, you may obtain a patent for a new and useful invention in such categories as the following:

- A process (primarily industrial or technical processes).
- A machine (like a bicycle).
- A manufactured article (like a clockwork radio).
- A composition of matter (for example, chemical compositions).

You can't obtain a patent for a mere idea, concept, or suggestion. You need to give a full and detailed description of the actual item or process when you apply for a patent. Also, you can't get a patent for an invention if someone else knew of, used, or described the invention in a printed publication *anywhere in the world* before the date you apply for a patent.

Registered designs

A registered design is a legal right with many similarities to patent rights but instead of focusing on how a product works it covers the product's visual appearance or 'appeal to the eye'. A registered design protects the product's shape or the pattern or ornamentation it carries. For example, you can protect the design of a chair, handbag, computer screen, or pen with a new and unique shape, or curtains or wallpaper carrying a new pattern.

The office that registers designs in the UK is the Designs Registry of the UK-IPO and in Ireland is the Irish Patents Office. For registered designs covering the whole of the European Union (including the UK and Ireland) the office is the EU Office for Harmonisation in the Internal Market (OHIM). You can find contact details for UK-IPO, the Irish Patents Office, and OHIM office, in Appendix B.

The application for design registration must include accurate and complete drawings of the product, along with the required formal paperwork and the filing fee (see Chapter 7 for more information).

The UK, Irish, and EU offices award registration for 5 years from the date you apply for it and you can renew for further 5-year periods up to a maximum of 25 years. The period of design protection elsewhere may be different, a notable example being the United States, where the protection comes as a 'design patent' that lasts for 14 years from the date the US Patent & Trademark Office grants it.

A registered design gives you the right (in the country of the registration and while it remains in force) to exclude others from making, using, selling, or importing articles with the design, to take court action to sue others for infringing the design (utilising it without your permission), and to license other people to use the design with your permission.

If your product not only looks good but also offers a new and useful function, consider seeking patent cover rather than design protection – or if you can afford it, apply for both. Often competitors can easily work around a registered design by changing the appearance of a product to make it 'uniquely different'. For example, your competitor may look at your square-backed chair and change the square to a more rounded back. This change may be enough to avoid infringing your design registration, and may itself qualify for a new registered design, but it probably still infringes your patent.

Registered trade marks

A *trade mark* is a sign, typically a word, phrase, logo, picture, or a combination of two or more of them, that identifies and distinguishes your products or services from those of others. Typical familiar examples for products are Dyson®, Rolls Royce®, Weetabix®, and Nokia®, and for services, BUPA®, McDonald's®, and Travelodge®.

A *registered* trade mark is a legal right that protects your mark against unauthorised use by someone else. The offices that register trade marks are similar or the same as for registered designs: in the UK the office is the Trade Marks Registry of the UK-IPO and in Ireland is the Irish Patents Office. For registered trade marks covering the whole of the European Union (including the UK and Ireland) the office is the EU Office for Harmonisation in the Internal Market (OHIM). For their contact details see Appendix B.

As well as words and phrases, perhaps surprisingly you can register colours, shapes, or sounds as trade marks. The airline EasyJet has a trade mark registration for its shade of orange. The shape of the Toblerone chocolate bar is registered. Microsoft registered their boot-up sound, and Direct Line registered their insurance jingle. For a mark to be registrable it must be capable of being shown graphically: for example in writing, drawings or musical notation. In principle smells can also be registered but presenting them graphically is difficult.

As the owner of the registered trade mark you have the right (in the country of the registration and while it remains in force) to take court action to sue others who, without your permission, cause confusion by using the same or similar marks for the same or similar goods or services. And if you want you can license other people to use the mark – for a suitable fee.

If you register your trade mark (in any country) you can indicate by the ® symbol that the mark is your registered property. And even if you don't register your trade mark, you can still indicate to the public that you regard it as your property by placing the ™ symbol alongside it.

Crucially, and unlike any other kind of IP, you can keep a trade mark registration in force for ever – subject, of course, to payment of renewal fees. In most countries the initial registration is for ten years from the date of application, renewable every ten years thereafter.

The possibility of keeping your trade mark registration in force for ever makes your trade mark potentially the most valuable part of your IP holding – by a long way.

Registering a company name as a trade mark is different from registering it as an Internet domain name and different from registering your company as a trading entity (in the UK at Companies House). Just because it's a domain name or registered at Companies House does not mean that the name satisfies the requirements of the Trade Marks Registry. In choosing your company name, do try to find one that's likely to be registrable in all three respects: trade mark, domain name, and company.

Chapter 4 covers how to search Trade Mark Registry records to check if someone else registered the mark you'd like for similar goods or services. Checking if a name is registrable as an Internet domain name requires contact with the appropriate domain name registry (for example Nominet for .co.uk domain names). Whether the name is acceptable as a UK company name requires a check with Companies House (see Appendix B for contact details).

Passing Off

Imagine you have a valuable trade mark for a successful product and you never got around to registering it, or maybe the trade mark failed to meet one or more requirements of the Trade Marks Registry. And then you find a competitor marketing a similar product under a very similar mark. What can you do? Fortunately, you may find a possible safety net in the UK in the common law offence of Passing Off.

The principle of *Passing Off* is that no one may conduct his business so as to lead customers to mistake his products or services for those of someone else. You can take this common law action against the offending competitor. But such court action over an unregistered mark usually results in more difficult negotiations or more lengthy court proceedings than for a registered mark.

Passing Off operates a very wide sphere of influence and may prove useful in actions that involve little or no trade mark activity. One potentially helpful field arises where a competitor markets a similar product to yours in a very similar presentation (known as 'get up') that confuses a customer seeing the packaged product on a supermarket shelf into thinking that it originates from you. Trading standards officers and the courts take a dim view of competitors who use this tactic.

Copyright

Copyright is not a right to copy! Quite the reverse: Copyright is a legal right to prevent copying – and a pretty strong one. Most importantly, copyright exists automatically from the moment you create the work in question.

Copyright applies to just about any kind of creative work: literary (books, magazines, newspapers, brochures, and so on, and articles or sections within them); musical (sound recordings, sheet music, orchestral works, chart tracks, audio tapes, CDs, film scores, and so on), motion pictures (including videos and DVDs); computer software; and art (paintings, sketches, or drawings).

The copyright we cover in this book is mainly *industrial copyright,* of which the most usual examples are computer software, two-dimensional sketches and drawings of a product, and other aspects such as the content of instruction manuals and advertising leaflets.

The three-dimensional shape of a product deserves a special mention because in the UK and EU it carries its own provisions: automatic and free design rights that arise whether or not you seek a design registration, and whether or not the design possesses any artistic merit. (See the later section 'Unregistered design right'.)

The period of copyright protection may be anything from three years from creation (for EU copyright in designs) to the lifetime of an author, composer, or film director plus 70 years from their death (in the UK and some other countries for dramatic, musical, or literary works). In the UK copyright protects sound recordings for 50 years from the date of publication and broadcasts for 50 years from first transmission. (See also Chapter 9.)

Although offering substantial protection, copyright does place upon you a strong burden of proof. You have to convince the judge that you own the copyright, that the offending competitor's offering is sufficiently similar to yours, and most importantly that the competitor truly copied your work. If the competitor can show that he didn't know of your work and created his own work independently, then he didn't copy you, which means he didn't infringe your copyright.

Unregistered design right

Even in the absence of a design registration in the UK or EU you acquire an 'unregistered' design right by your creation of any original, non-commonplace designs of a product's shape or configuration. It protects both functional and aesthetic aspects, so you don't need to show the design to have artistic merit.

This unregistered design right is a type of copyright, giving you the right to take action against anyone copying the design. Like other kinds of copyright, but unlike a registered design, you must be able to prove that there was copying: the unregistered design right does not protect you against someone who can show that their similar or even identical design was not created by copying yours.

Another possible weakness of the unregistered design right is that it applies only to three-dimensional aspects of design. Unlike a registered design it cannot protect surface decoration or ornamentation. If these are the important elements of your design then you need to consider applying for a registered design. (See Chapter 7 for more on registering designs).

The period of protection for the UK unregistered design right is 15 years from the end of the year in which the design was first recorded, except that the period is reduced to 10 years if marketing of the design took place within 5 years of that first record.

For the first five years you can stop anyone copying your unregistered design. For the rest of the time period the design is subject to a *licence of right*, which means that the authorities entitle anyone to get a licence to make and sell products copying the design (in return for suitable payments to you).

As with other kinds of copyright you face a high burden of proof when trying to enforce your unregistered design rights. Keep good records to demonstrate your creative activities, most especially the date you first recorded the design in a design document.

Plant breeders' rights

Many countries grant rights to protect new plant varieties. In the UK they're known as *plant breeders' rights*, and they protect new varieties of plants or seeds. You can apply to register your new plant or seed varieties with the UK or EU Plant Variety offices. Unlike other kinds of IP in the UK, responsibility for the rights does not come under the Intellectual Property Office (UK-IPO) but the Department for the Environment, Food, and Rural Affairs (DEFRA) www.defra.gov.uk.

The UK Plant Variety Rights Office (PRVO) and the EU Community Plant Variety Office (CPVO) provide the forms that you need to apply for respectively UK and EU-wide plant breeders' rights. As the holder of such rights, you can prevent anyone producing, selling, importing, or exporting the plants or seeds. See Appendix B for PRVO and CPVO contact details.

You can protect rights to a plant variety in the UK or EU if you can clearly distinguish it from any other variety in common knowledge. In this context, *common knowledge* includes a variety that's already in cultivation, or has been commercially exploited, or falls under a plant variety right in another country.

The PRVO and CPVO grant plant breeders' rights for a term of 30 or 25 years, depending on the species. In all cases, the rights run from the date of grant and are subject to the payment of all relevant fees during the term.

In the US plant protection comes in the form of a *plant patent*. Anyone who invents, discovers, or sexually reproduces any distinct and new variety of plant can apply for a plant patent. The plant patent lasts for a term of 20 years from granting and requires no maintenance fees.

You may also be able to register the name of your new variety of plant or seed as a trade mark. See the section 'Registered trade marks', earlier in this chapter, and Chapter 8 for how to register a trade mark.

Know-How and Trade Secrets

Know-how and trade secrets aren't among your formally published or granted legal rights, and indeed you wouldn't want them to be. But know-how and trade secrets can nevertheless be of vital importance to the success of your business.

Know-how is the practical knowledge, skill, and expertise that arises as you go about your business: the little actions that make things run smoothly and efficiently, the ability to avoid wasted effort, and generally knowing how to make things work well.

A *trade secret* is information that you want to protect and keep secret to give yourself an edge over your competitors. Whether this information is a design, process, formula, composition, or technique, it can give you or your company useful advantages.

When you get round to licensing your IP rights to third parties it can often be the detailed know-how and secrets associated with those rights that enable you to put a high price on the deal. So in the meantime keep these nuggets of wisdom to yourself. You don't want competitors to discover from you the tricks of the trade that make your products that bit better and easier to make.

Often your trade secrets embody patentable information. You have to decide whether you want to obtain a patent and have time-limited protection, or to take steps to keep the secret to yourself for an unlimited time. This section helps you decide whether and how to protect your know-how and trade secrets.

Looking at examples of trade secrets

The more common trade secrets include recipes and processes, such as the formulae for food, soft drinks, cosmetics, and chemicals. One famous trade secret is the formula for the Coca-Cola® brand of cola drink. If Coca-Cola had patented the formula it would have become public information. But Coca-Cola chose not to patent its product, so even though the can or bottle contains an ingredient lists, no-one else knows how much of each ingredient to put in, nor the method of combining them.

You can consider business methods and techniques as trade secrets. Such methods and techniques may include your customer mailing lists, pricing and distribution techniques, and warehousing turnover. All these may give your business a competitive edge. Look after them!

Protecting know-how and trade secrets

Do think about and implement precautions to maintain the confidentiality of your know-how and trade secrets. If you need to disclose your secret to third parties, have them sign a confidentiality agreement undertaking not to disclose the secret to anyone else (see Appendix A for a specimen wording, or use the version on the UK-IPO website – contact details in Appendix B). If visitors to your premises may encounter secrets, get them to sign in and confirm in writing to respect the confidentiality of what they see and find out there. And escort such visitors on the premises at all times until they leave.

Preventing employees from spilling the beans

Get your employees to sign a written contract of employment and include in the contract your terms of keeping your pertinent business information confidential. If an employee breaks this agreement and reveals confidential information you can sue for breach of contract. For more on this issue, see Chapter 11.

When an employee leaves, especially one who knows much of the firm's private information, conduct an exit interview to remind him or her of their obligation to keep your information confidential.

If a competitor is hiring the departing employee, write to the competitor to tell them that the employee is under an obligation not to disclose your secrets. Some companies send the competitor a copy of the employment contract or confidentiality undertaking that the employee signed.

Examining pros and cons of trade secrets

A trade secret has several positive attributes. It doesn't involve the time constraints of a patent, which expires after a set period – the trade secret can in principle remain yours alone forever. Having a trade secret avoids the time and costs incurred in getting a patent. And getting around a trade secret is harder than for a patent, because others don't know what the secret is!

However, a trade secret also has disadvantages. No matter how hard you try to keep it secret this may prove impossible. Competitors can copy or even reverse engineer many things, however sophisticated the product. Or another person may legitimately discover the secret, may be able to patent it himself, and then sue you for patent infringement. You may be able to overcome this by proving your prior use – in which case the authorities allow you to continue your use without suffering any penalties – but the proof is likely to involve long and costly legal proceedings, and you lose your exclusive position.

International Rights

Your first thoughts on getting IP rights may go no further than filing a patent application in your home country, possibly along with design and trade mark applications. This section reminds you a bigger world exists out there and here we guide you on how to extend your protection into other countries.

A UK patent gives you a monopoly in selling and distributing your product, but only in the UK. Unless you gain equivalent protection in another country, anyone can make, use, and sell your invention in that country without paying you a penny.

Make sure that a commercial market for your product exists in a country, or strong indications show that the market is going to exist, before spending a lot of time and money on IP rights there.

Nearly every country operates its own intellectual property laws and the offices and officials to operate them. But nearly every country also belongs to one or more of the international conventions, treaties, protocols, and agreements developed over the years to assist with gaining IP rights. Government offices have also made immense efforts to harmonise IP laws and procedures to make them conform more closely between countries. The following sections review the international help on offer.

Avoiding Negative Meanings

Be very careful of the trade marks and words on the packaging you use when selling abroad. Vocabularies differ and various words may have negative or derogatory meanings in other countries.

In finding a name for the replacement for their Silver Cloud car, Rolls Royce naturally looked for

another 'Silver' and came up with Silver Mist. Fortunately, someone told them what 'Mist' can mean in German. It was at once clear that their discerning German customers would not have been keen to drive around in a Silver Poo. The car became the Silver Shadow.

Many of the international IP arrangements offer significant savings in costs and management time because a single set of procedures covers several countries, saving you from making individual applications in each country. A large part of the saving comes from avoiding or deferring the costs of translating the application papers for every country with a different language.

The Paris Convention

The Paris Convention, dating from 1883, was one of the first international agreements on IP. Just about every country in the world is a member (172 contracting states in 2007), with the notable exception of Taiwan, which remains excluded as a result of political differences with mainland China.

The convention's most used provision allows the date on which you applied for IP rights in your home country to be treated as the priority date of a later but equivalent application in another member country. For patents, you must make the further application within a period of 12 months from filing the earlier application. For registered designs and trade marks, the period is six months.

The value of this provision is that your home application gets the details of the IP officially recorded at a date that can be recognised in the equivalent applications worldwide. So in a single application you've laid the basis for protection in most other countries around the world.

The Paris Convention gives you a few months after filing your application in your home country to reflect, and possibly test the market, before making a decision – and incurring the costs – for equivalent applications elsewhere.

The Patent Cooperation Treaty (PCT)

The PCT system lets you file a single international patent application – in English and filed initially at your home IP office – that designates any of the PCT member states. The member states (137 in 2007) include most countries you've ever heard of, again excepting Taiwan.

By using the PCT you face the early stages of the patenting procedure only once for all the countries of interest to you. The early patenting stages include publication of your application, a search for any prior publications of similar inventions to yours, and optionally a preliminary examination to give a reasoned opinion on the likelihood of gaining patent protection.

The body responsible for PCT applications is the World Intellectual Property Office (WIPO), based in Geneva. However, WIPO delegates the search and examination activities to other IP offices. The European Patent Office (EPO) undertakes PCT searches and examinations for applicants in the UK and other EU countries. (See Appendix B for contact details of WIPO and EPO offices.)

The PCT system does not directly grant patents. After the PCT stages, which can take up to two and a half years to work through, you continue with patent applications in each of the countries – or maybe regions (see below for details of 'regions') – that you're still interested in. You probably only make such patent applications if the PCT stages give indications of getting strong patent cover and if the market prospects continue to look good.

Regional patent systems

Countries in several regions have joined together to create patent systems in which a single application can take you through at least some of the stages towards a granted patent. Patent systems operate in a similar manner to the PCT system but across the region rather than worldwide, and unlike with the PCT, the granted regional patent can be validated without further examination for it to become effective in the required countries. (An application emerging from the PCT system must go through further examination in the chosen region or country.)

Current (2007) regional systems include the European Patent Convention (32 countries in Europe), the Eurasian Patent Organisation (9 former Soviet countries), the African Regional Intellectual Property Office (ARIPO; 16 countries, mostly in East Africa), and the Organisation Africaine de la Propriété Industrielle (OAPI; 16 countries, mostly in French-speaking Africa). We give contact details for all four in Appendix B.

The European Patent Convention (EPC)

You can file an EPC patent application in English (or in French or German), designating any or all the member countries. The member countries include all the members of the EU, plus Iceland, Monaco, Switzerland, Leichtenstein, and Turkey. A notable absentee is Norway (although Norway is a member of the PCT).

The European Patent Office (EPO) processes your filed application as a single application. The processing goes further than that of the PCT. After publication and a search stage the application undergoes a full examination. If the EPO accepts the application, it grants you a European patent, which you can validate as separate patents in your designated countries. If you ever need to, you can take court action to enforce the patents by proceedings in the respective countries' courts.

The European Union Patent? (if only)

Despite its name, the 'European Patent' that you obtain from the EPC takes effect in individually designated countries rather than the EU as a whole (and in designated EPC but non-EU countries such as Switzerland).

Countries have made many attempts to create a system that grants a single patent effective across the EU and that a single European IP court system enforces. But agreement among the member states has proved elusive, due to arguments over the official language(s) and the provisions for the courts. So at the publication date of this book no EU patent exists.

European Union Designs

The European Union Design System provides for uniform protection of designs across the EU. The protection is of two types: the Unregistered Community Design and the Registered Community Design.

Unregistered Community Design Rights

The Unregistered Community Design is similar to the UK unregistered design right (which we describe in the section 'Unregistered design right', earlier in this chapter). This right is also automatic, free, and a right against copying, but its term of protection is shorter at just 3 years from the date of first publication in the EU (compared with 10 or 15 years' protection in the UK). You need to keep documentary evidence of the creation of the design in case it becomes necessary to take legal action to enforce your rights.

Registered Community Designs

You can seek a design registration for the Registered Community Design (RCD) for the whole of the EU by a single application. The Office for the Harmonisation in the Internal Market (OHIM) – an Alicante-based organisation responsible for granting EU design and trade mark registrations – processes the application.

You can file the application at OHIM or your EU home country IP office, but the latter is likely to incur a handling fee. Chapter 7 goes into more detail on the filing needs.

No official searches or novelty examinations of the application take place for this application, but OHIM do check for basic formal requirements, such as including adequate drawings or photographs.

Registration and publication of the design typically occurs about three months after filing. The registration term is an initial 5 years from the application date, renewable for 5-year periods up to a maximum of 25 years.

We strongly recommend that you consider applying for EU registration of your designs. If you have market interests in two or more EU countries, this is likely to be the least expensive route to registration. Not only do you get protection across the whole EU, but you also benefit from the wide range of protection that can extend to many signs and symbols that also qualify for trade mark registration. And whereas for a trade mark you may incur the registration fees in several classes of goods, the single design registration covers them all.

The Madrid System for trade marks

The Madrid System for registration of trade marks includes the Madrid Agreement and the Madrid Protocol. The protocol is the more recent and in several respects the more user-friendly. If you own a trade mark in a member country of the agreement or protocol you can protect it by an international application that covers other member countries – a total membership of 80 countries. Further information on the system is given in Chapter 8.

You can file the application at your home country IP office (in the UK, the UK-IPO), which then forwards your application to WIPO for processing.

The international registration lasts for ten years from its date of registration and you can renew it indefinitely for further ten-year periods, subject to the payment of renewal fees.

European Union trade marks

You can register a trade mark that covers the whole of the EU (generally known as a *Community Trade Mark* – CTM) through a single application to OHIM, following very similar processing to that for a Registered Community Design (see the earlier section on the RCD under 'European Union Designs').

As with the RCD, you can file your application at OHIM or at your EU home country IP office, but the latter is likely to incur a handling fee. See Chapter 8 for more on the Community Trade Mark.

If you're to register the mark, it must be distinctive and free from conflict with the prior rights of others. OHIM examines the application to check that it's formally correct and undertakes a search for prior registrations. OHIM's search report sent to the applicant indicates earlier identical or similar CTMs and CTM applications for identical and similar classes of goods and services, together with national search reports from most of the member countries.

A CTM registration lasts for ten years from its date of registration and you can renew it indefinitely for further ten-year periods, subject to the payment of renewal fees.

The CTM system offers a very cost-effective way of protecting your trade marks in Europe. In general, if your market interests are in two or more EU countries, the CTM route is the one to consider.

International copyright conventions

International conventions on copyright ensure that when nationals or residents of a member country create copyright material, another member country protects that copyright by the national law of that country. The periods of protection for different countries vary slightly but are generally similar.

Over 160 member countries have adopted the Berne Convention on copyright, which comes under the responsibility of WIPO. Another major convention is the Universal Copyright Convention (65 member countries), which falls within the cultural sector of the United Nations Educational, Scientific, and Cultural Organisation (UNESCO).

The Rome Convention, also administered by WIPO and with 75 member countries, relates specifically to the copyright in 'phonograms', meaning sound recordings on a base such as a CD or audio tape, but not the sound tracks of films, DVDs, or video cassettes.

International plant breeders' rights

The International Union for the Protection of New Varieties of Plants (UPOV) is a 64-member intergovernmental organisation, based in Geneva, established under an international convention for the protection of new plant varieties. WIPO provides its administrative services. See Appendix B for contact details.

World Trade Organisation's obligations

The World Trade Organisation (WTO) has obligations to protect IP under an international agreement on the 'Trade-Related Aspects of Intellectual Property Rights' (TRIPS).

TRIPS covers such issues as how countries apply the basic principles of the trading system and other international IP agreements, how to give adequate protection to IP rights, how countries enforce those rights in their own territories, and how to settle IP disputes between members of the WTO.

As with so many other international aspects of IP, you probably never need to encounter TRIPS directly. You may, however, be reassured by the extent of support for your IP this and the other international arrangements offer. The recognition by WTO of the importance of IP is particularly welcome.

Hiring an IP Professional

The laws and procedures of the IP world are complicated and full of potential pitfalls. You may feel able to take on yourself a number of the early steps for checking out other people's rights and perhaps in filing your first applications for protection, but sooner or later you're likely to appreciate a professional guide. As highly qualified specialists, IP professionals' services don't come cheap. However, their experience and expertise can be invaluable.

This section discusses the different types of IP professionals and how they can help you through the application and registration processes. Chapter 4 also provides information on using professionals.

To make yourself aware of the probably substantial costs of using professional IP help, we advise you to get an indication in advance of the charges incurred in a given activity. Most IP attorneys provide a free first hour of consultation with you, in which they advise on the kinds of protection that may be available and give an estimate of the costs involved.

The specific qualifications of IP attorneys and the forums and courts in which they may represent you vary from country to country. The following sections apply mainly to the situation in the UK.

Patent attorneys

A patent attorney is generally someone with a background and qualifications in such areas as engineering, chemistry, biology, and information technology and who qualified as an attorney through the demanding exams administered by the responsible official body (in the UK, by the Chartered Institute of Patent Attorneys).

A strict code of professional conduct binds patent attorneys, especially with regard to keeping your information confidential. This means that you can reveal the fine detail of your invention to your patent attorney without fear that he'll take it and exploit it himself. And you need to reveal the fine detail to make sure that the advice he gives and any patent specification he prepares leads you to the maximum available protection.

Most UK patent attorneys also have experience in dealing with design issues and industrial copyright, and are also qualified as *European patent attorneys*. They can represent you not only at the UK Intellectual Property Office but also at the European Patent Office and at certain courts, notably the UK Patents County Court.

When it comes to representation in the UK High Court you usually need to appoint, in addition to your IP attorney, both specialist IP solicitors to assist in preparing the court evidence and documents and specialist IP barristers to present your case in the courtroom. Using professionals is expensive, but fortunately only a handful of cases ever get into court. You can try many stages of negotiation or mediation – and one of them usually succeeds – before reaching the court door.

Patent attorneys are *not* qualified to give a marketability opinion of your product, only a patentability opinion. They are two very different areas.

Trade mark attorneys

Trade mark attorneys deal with searches and applications involved in registering trade marks. They face professional qualifying examinations (the Institute of Trade Mark Attorneys administers the examinations in the UK). Most IP firms include both patent attorneys and trade mark attorneys (sometimes the same people are qualified as both) so as to provide the most appropriate professional help for your needs.

Try to find an IP attorney through personal referral from another inventor, a friend, an inventors' organisation, or your other professional advisers such as your accountant. You can also go to professional bodies such as the Chartered Institute of Patent Attorneys or the Institute of Trade Mark Attorneys for lists of their members (we provide contact details for both bodies in Appendix B).

IP solicitors

Solicitors in the UK may also represent you in IP matters. But do be careful if you look to small local firms of solicitors for IP help as they may have limited experience in such matters. At the other end of the scale, specialist firms in major cities have vast IP experience and expertise, in particular with regard to enforcing IP rights in the courts. All UK solicitors are bound by the requirements and standards of their professional body, the Law Society (we provide contact details in Appendix B). If you ever face High Court action – and we hope you never do – both your IP attorney and IP solicitor may recommend an IP barrister with suitable qualities and experience to present your case.

Chapter 4

Searching for Patents, Designs, and Trade Marks

*Y*our ideas may be new to you, but you don't know if they're truly new until you do a search. Other people in the UK and overseas already hold loads of intellectual property rights, publish or patent millions of inventions (although most never see the light of day as a product), register lots of designs, and grab many, many trade marks.

Until you conduct a search, you've no way of knowing whether you may infringe upon another's rights. And you don't want to wait until the official search, after all your efforts in preparing and filing your application, to find out that your idea isn't unique.

Although you don't have to conduct a search before submitting an application for a patent or other intellectual property rights, we strongly recommend that you do . In fact, even if you decide to hire a professional to conduct a search for you, think about conducting one yourself, first. By doing so, you discover more about the intellectual property law process, similarities with other inventors' ideas, and, possibly, new commercial applications. And conducting your own search can prevent you from pawning the family jewels only to find that what you thought was a new, patentable idea was actually invented by the cave men.

This chapter explains how to determine what's new about your idea, how a search can help improve your idea, how to prepare and start the search, and how to enlist the help of the pros. It also discusses how to use your search results.

Do I Really Need to Do a Search?

You may say, 'Well, I've looked everywhere. I've been to Tesco, John Lewis, and B&Q and I've been through the Argos catalogue, and I haven't seen my product anywhere. My product's not patented. I don't need to conduct a search.'

Don't assume any such thing. Just because your product isn't on the shop shelves doesn't mean that someone doesn't hold a patent or registered design for your product.

Remember that no more than about 5 per cent of patented products get on the market. The other products don't, so you won't find them anywhere but in a patent search. Likewise, that rather appealing product shape you're so pleased with may be registered to a creative artist from Padstow. You don't want to work day and night to establish a small business around your product and then find out that someone else already holds the rights to it.

Remember too that someone else may have registered the word or logo you want to use as the brand to attract buyers, for the same or a similar product to yours. Here a search of the official register of trade marks can avoid the risk of finding out later that the brand was not yours to use and so forcing you to choose a different one. If you don't check the official register until near to your product launch you risk the unwanted costs of relabelling or reprinting the product, packaging or publicity. (See Chapter 8 for more on trade marks and how to protect them.)

Standing on the shoulders of giants

One of the greatest scientists and mathematicians of all time, Sir Isaac Newton (1642–1727), derived the laws of gravity, invented mathematical calculus, and created a vast range of innovations in chemistry and physics. He developed many of his ideas from the writings of English physicists Robert Boyle and Robert Hooke, and the French scientist René Descartes.

When Robert Hooke congratulated Newton on the extent of his achievements, Newton wrote in reply: 'If I have seen further it is by standing on the shoulders of giants.' Even this quote was not entirely original: Authors throughout the middle ages recorded versions of the phrase. As early as 1190 AD one Bernard of Chartres observed, 'We are like dwarves on the shoulders of giants, so that we can see more clearly than they.'

The lesson for later inventors is clear and often put more simply: Don't reinvent the wheel! First study what others have already done and then invent something better.

Large corporations conduct searches in their given areas of interest to discover enticing potential improvements, applications, and alternatives that they can take advantage of in the manufacturing and marketing process. A wise investor usually wants to know whether you have a patent or other intellectual property rights. If you don't have any IP rights, the investor usually asks if you've conducted searches. If you seek investment money but don't do searches, corporations generally run away from you. Do your homework! (See Chapter 22 for more on negotiating with corporations.)

Read the information on the Web site of the UK Intellectual Property Office (www.ipo.gov.uk), which explains how to conduct a search for patents and other kinds of intellectual property and walks you through the step-by-step process.

Determining Whether Your Idea Is Really New

After you find a way to demonstrate to your own satisfaction that your invention works and does what you think it does, you need to carefully decide what's really new about it – its *novelty*. Until you identify what your novelty is, you can't conduct a focused search to find out how your invention differs and improves on what's already available to the public. This doesn't just mean it must have novelty over existing products or processes, it needs also to have novelty over all information already available to the public in any way (written or spoken) anywhere in the world.

In addition, your invention needs to show an improvement, usually termed the *inventive step,* over what's already available to the public. Otherwise why would anyone want to buy your invention and how can a patent office feel justified in granting you exclusive rights to it?

Suppose that you invent a bicycle tyre. Think about the specific features that you invent to make the tyre novel or unique. The new features are what you may be able to patent. Take a close look at all the similar patents to see whether the features, functions, and novelty of your invention really are unique. And concentrate on what makes your tyre better.

Searching for Patents

By conducting a patent search you become more knowledgeable about your product and its competition. Not only can you can see if similar patented products to yours exist, but you also take in a significant amount of information by reviewing other inventors' patents and published applications.

You see the drawings, legal descriptions, technical information, and, possibly, the operational components of the inventions. You may not have considered some of the ideas expressed in others' specifications, and you can use those insights to improve the overall design of your product.

A patent search can enlighten you about additional commercial uses for your product. You may also find that the major breakthrough you think that you have is only a small improvement on what's already available. Knowing this can help you better position your product when it comes on the market. (See Chapters 19 and 20 for more on marketing your product.)

Your patent search can also help you write your own patent specification. Reading how other people describe their inventions in their patent specifications guides you on the style and contents typically required. Chapter 5 includes a detailed section on what a patent specification should typically include.

What if you don't feel happy about doing your own search (though simple early searches can be quite straightforward – see the 'Utilising the Internet' section, later in this chapter) or you feel that your initial search hasn't gone far enough? Then you can hire a professional searcher or attorney to do or extend the search. See the later section 'Getting Professional Help' for how to go about this.

You can conduct your own patent search over the Internet, with help from a whole range of user-friendly databases. A selection of these databases is given in Appendix B.

Preparing for an *ad*venture

Although some people refer to inventing as a venture, the ups and downs of inventing make it more of an *ad*venture in understanding life and the realities of a business. Most successful inventors go through the typical pain and misery associated with taking a significant amount of risk, spending long, hard hours dedicated to testing and improving their ideas, and enduring a shortage of available resources. Just get ready to put your boots on and come out of the corner fighting if you decide to take the typical road toward an adventure.

But be reassured that you're not the first to follow the long adventure road through to success. Georges de Mestral worked for years on the mechanics of plant burrs before arriving at reliable designs and materials to create the Velcro fastener. James Dyson built over 5,000 prototypes until he reached a design that reliably applied the dust-removing properties of cyclones to a domestic vacuum cleaner. Turn to Chapter 25 for more on their great success stories.

Secrets of the America's Cup

The America's Cup takes its name from its first holder, the New York Yacht Club's (NYYC's) schooner *America*, which easily beat 14 British boats in the inaugural races in 1851 around the Isle of Wight. Queen Victoria watched the races and predictably was not amused. Challenges for well over 100 years failed to wrest the trophy from the NYYC, but gave the impetus for huge innovative advances in vessel and sail design from defenders and challengers alike.

In 1983 the impressive early performance of an Australian challenger, *Australia II*, attracted much interest. The press and the yachting public suspected that the reason for the boat's success was its radical new keel, developed by designer Benny Lexcen. The Australians took care to hide the keel between races by wrapping a green canvas skirt around it, which increased public curiosity as to what lay beneath. *Australia II* continued its winning streak and duly lifted the trophy on 26 September 1983, ending 132 years of NYYC dominance, with its keel secrets still intact.

Or were they? Benny Lexcen had applied in 1982 for patent protection for his inventive keel. A patent search before the races started would have found his British patent specification, published on 24 August 1983 with the number GB2114515. The specification reveals that his keel had small side fins, discloses the angles at which they were attached, and includes drawings showing their precise location.

It seems the press and the yachting types were unaware of how much you can discover from a patent search. They could have known the keel's secrets a month before the crucial race.

The European Patent Office (www.epo.org) has developed one of the best resources – Esp@cenet database (ep.espacenet.com) – and it's free. This database lets you fine-tune your search by the following:

- ✔ Entering *keywords* (words and phrases that the individual titles or abstracts in a patent contain).
- ✔ Calling up an individual patent specification if you already know its official number.
- ✔ Using the classification system to find all the patent specifications within a chosen technical subject heading.

A good route to Esp@cenet is via the Web site of the UK Intellectual Property Office (www.ipo.gov.uk) – follow the links though 'Patents' and 'Find Patents').

Esp@cenet provides worldwide coverage, including patent specifications from the UK, Europe, the USA, Canada, Japan, China, Korea, and just about everywhere else. And the records go back 100 years and more. In most cases the search provides an English language title and abstract of the patent specifications it finds, and lets you inspect or download the whole specification.

The Esp@cenet keyword searches scan patent titles and abstracts but not the full descriptions. If you want a search that can scan whole specifications – for example, if you're looking for prior use of 'titanium' but the Esp@cenet-revealed titles or abstracts just mention 'metal' in general – then try the 'whole specification' search option on the Software for Intellectual Property Web site (www.patentfamily.de).

To search the Web sites of other countries, try looking first at the UK Intellectual Property Office Web site. It includes a full listing of overseas Intellectual Property Offices and their contact details (click on 'IP Abroad').

Searching by subject

Some basic steps for a patent search by subject include:

- ✔ **Search by keyword.** Your search is only as good as your selection of the keywords in the first place. Include keywords of the features of your invention that you think are important. For example, say you're working on improved brakes for a baby buggy. Try keywords such as 'baby', 'brake', 'buggy', or 'carriage'. Esp@cenet allows you up to four keywords at a time and quickly lists all the patent specifications containing your selection. If the keywords 'baby', 'buggy', or 'brake' produce hundreds or thousands of hits (and you sensibly don't want to look at all of them), try limiting the search to 'baby buggy brake'.

- ✔ **Use a good existing patent.** A good patent specification to help your search is one with features closely resembling what you're trying to patent. That way, you know that you're in the correct area. When you conduct a search, if no similar patent specifications exist, determine whether you're really conducting the search in the correct area. Similar ones generally appear.

  Enter the specification number of the good patent into the Esp@cenet 'number search' box. This brings up the title and abstract and shows you the class and subclass where the invention is primarily referenced and cross-referenced. Esp@cenet lets you inspect the full description, claims, and any drawings, and often the search list of prior art found by the Patent Office examiner during the application stage. Prior art simply means products for sale, patents, published pending applications, or newspaper and other written information that's available to the public.

- ✔ **Search by classification heading.** Patent Office examiners classify patent specifications into detailed classification systems, grouping those with similar subject matter into the same classes or subclasses, each with a heading number. You can find the European classification headings via Esp@cenet – on the UK Intellectual Property Office Web site go to 'patents', 'find patents', Esp@cenet, and 'classification search'. Enter your keywords and see which numbered headings they produce.

The European classification system puts 'baby buggy' into heading B67B7 'carriages for children'. A combined search of B67B7 and 'brake' (using the Esp@cenet 'advanced search' option) produces a list of 30 or so patent specifications meeting this combination.

Getting all the details

Esp@cenet and similar sites provide information on the patent specification's title, abstract, description, claims, and any drawings; the names of the inventors and applicants; the filing dates and numbers, and whether the patent is in force. You may view and print any or all this information for free.

You usually conduct patent searches to help you determine the patentable and marketable features of your invention over what's already known, rather than to discover whether your invention falls within the claims of a wider earlier patent that's still in force. If you find an earlier patent that your invention may link to, seek professional advice to find out if marketing your product – even if inventive in itself – runs any risk of infringing that patent.

If your own searches didn't find any similar patents, think about whether you did the searching correctly. Millions of patent specifications exist and usually something at least vaguely similar turns up. You may have been looking in the wrong place! Go back and approach the search from a new angle, or think about getting some professional help.

Searching for Registered Designs

A search for existing registered designs (which we discuss in detail in Chapter 6) is usually less important than a patent search. However, if the marketability of your product lies mainly in the visual appeal of its shape or surface features (rather than in the functional details, as in a patent as such), we strongly recommend that you do a design search, even if you have to pay a professional to do it.

Searching for existing registered designs is very like searching for patents. If you don't feel confident or competent to do a design search yourself, enlist the help of a professional searcher. Flick forward to the 'Getting Professional Help' section, later in the chapter, for the lowdown on hiring a professional.

But you may be pleasantly surprised by how easy doing your own preliminary online search can be. For UK registered designs, go to the Web site of the UK Intellectual Property Office (www.ipo.gov.uk) and follow the links though 'designs' and 'find designs'. You can search their records free of charge using the design's number, product, proprietor, or classification.

For European 'Community' designs, remember that European Union design registration covers all EU member states, including the UK. Go to the EU office's Web site (www.oami.europa.eu); click on 'OHIM' – to give the English language version – then on 'community design', 'databases', and 'RCD online'. Next put your field of interest, for example 'bicycle', into the 'indication of product' box.

For US designs, go to the Web site of the United States Patent and Trademark Office (USPTO; www.uspto.gov). Click on 'patents', 'patents search', and then 'quick search' or 'advanced search'. Because the USPTO award design protection as a *design patent* – which differs slightly in scope, term, and procedures from a registered design in other countries – you use the same searching route as for *functional patents*, which the US define as *utility patents*. (See Chapter 3 for more on US utility and design patents.)

If you're interested in designs in other countries, try the Web sites of their IP offices. We list the most frequently sought Web sites of other countries' IP offices in Appendix B. You can also find a full listing on the UK-IPO site (www.ipo. gov.uk – click on 'IP abroad'). But depending on the country and its local language, you may find you need professional help to be sure of getting the most useful details.

Searching for Trade Marks

In the long run the trade mark applied to your product or process may be the most valuable element of your intellectual property. So putting some effort into choosing one, especially searching for one that someone else hasn't already registered, is well worthwhile. Have a look at Chapter 8 for more on trade marks.

Use a professional searcher if you don't feel confident to do the search yourself. However, the routes are similar to those for patents and registered designs (see the earlier sections in this chapter) and are user friendly, so consider giving them a try, at least for your preliminary searches.

For UK-registered trade marks the first route is again via the UK Intellectual Property Office (UK-IPO) (www.ipo.gov.uk). Follow the links though 'Trade Marks' and 'Find Trade Marks'. You can search for existing marks by number, proprietor, image, text (the letters, words and/or numerals used in the mark – a likely field for a preliminary search), or by classification.

Helpfully, the UK records also include European Community trade mark registrations (which cover all member states of the European Union, including the UK). But if you want to check out the EU records as such, go to the EU office Web site (www.oami.europa.eu), click on 'OHIM' – the English language version – then on 'Community Trade Mark', 'databases', and 'CTM online'.

For US trade marks, go to the USPTO Web site(www.uspto.gov). Click on 'Trademarks' (written as one word in the USA) and then 'Search TM database (TESS)'. As a new user, go to the 'New User Form Search (Basic)' and enter your proposed mark in the 'search term' box.

To find registered trade marks from the official Web sites of other countries, look at those we list in Appendix B or the full listing on the UK-IPO site (www. ipo.gov.uk – click on 'IP abroad'). Depending on the country and its local language, however, you may need professional help to achieve a result you feel able to trust.

Getting Professional Help

This section discusses how to go about getting expert advice, as well as what to expect and look for when you select your professional helper.

Aside from any of the preliminary searches you make to improve your awareness in your field of activity, you're best asking a professional to conduct the main searches for patents, designs, and trade marks. A professional knows the ins and outs of the system and can think of places to look that you may miss. Their searches may find prior art (see the earlier section 'Searching by subject' to find out what this is) similar to what you believe to be new about your project; may show that your idea is an improvement on the invention of another; or may find an existing in-force trade mark registration for the mark you hope to use.

For help with your searching you've a choice between using professional searchers and using a patent or trade mark attorney. Professional searchers are experienced in searching but, unlike attorneys, aren't usually qualified to deal with such aspects as how to define your invention in legal terms or if your invention infringes any of the rights found in the search. In the early stages of developing your invention or choosing a trade mark you may decide that an established searching firm can adequately meet your needs. We give you a non-exhaustive list of searching firms in Appendix B. Their charges are likely to be less than if you go directly to a highly qualified attorney.

As your ideas develop and your awareness of prior art and prior rights grows, perhaps helped by your own searching efforts, you may prefer to go directly to an attorney. Any searching conducted by the attorney can then be first stage of the process of preparing your patent, design, or trade mark application, and perhaps ultimately of having to defend your granted rights against infringers.

Your patent, design, or trade mark is only as good you can defend it in a court of law. If your product is really good someone is quite likely to infringe it, so it may only be a matter of time until you're in court to enforce your rights.

When selecting an attorney, get references from local inventor organisations (some of which we list in Appendix B), other professionals, and friends. One of the main things to look for when you select a patent attorney is to make sure that their background is similar to that of your invention. For more on hiring professional help with intellectual property, refer also to Chapter 3.

Questions to ask

As a novice inventor, you may not know the kinds of lawyers to approach or the revealing questions to ask them. Patent attorneys deal with patent or registered design applications or issues of industrial copyright. Trade mark attorneys deal with trade marks. Some individuals are qualified as both patent and trade mark attorneys but tend to concentrate on one or the other.

Here are some questions to put to a prospective patent or trade mark attorney.

- ✔ **Can I patent or register my product?** You may not receive a direct 'yes or no' answer, but merely an opinion concerning the likelihood of success. Then you'll have to decide if the probability of receiving protection is great enough to justify the risk of investing time and money. No guarantees exist. You don't want to do all the work necessary to make money from your product if you don't have the possibility of getting worthwhile protection.

- ✔ **How many patents (designs, trade marks) exist that are similar to mine?** Ask this question after the initial search. Then, based on the search results, ask again whether you can patent or register your product. You may be surprised by the answer, especially if a large number of patents are similar to yours. If several closely similar patents exist you may want to think twice before starting your venture. Their inventors have already taken a forward step. If you don't see any of the products from similar patents for sale, is there really a market for yours?

- ✔ **What is your experience?** How many applications have you filed? Have you been involved in any infringement cases? How long have you been practising?

- ✔ **What is your area of expertise?** (You ask this question primarily when seeking a patent attorney.) What's your degree in – mechanical, electrical, or environmental engineering, or biology or chemistry, for example? Remember to select your patent attorney by the type of product you want to protect.

- ✔ **How do you bill your clients?** Do you charge an hourly rate? If so, how much? Some attorneys require a significant amount of money upfront, whereas others bill by the pay-as-you-go method.

- ✔ **How do you handle renewal fees?** Do you send out reminders when a renewal fee is due? How many reminders? How far in advance of the due date? Do you delegate renewal matters to an outside agency?

    The need to pay regular renewal fees to keep your rights in force is a painful part of owning intellectual property. (See Chapter 6 for more on patent renewal fees and procedures.) Attorneys operate their own renewals department or outsource the task to a specialist renewal agency.

Knowing the cost

If you conduct the search yourself, the information's free for the asking. If you hire a professional searcher, the cost may be hundreds of pounds, as may the services of an attorney – who almost inevitably costs more than the searcher (after all, they're more highly qualified). All these trained professionals are, however, well worth the investment. When conducting your own search, you may miss some prior art (see the earlier section 'Searching by subject' for a definition) from the area you search. But before you instruct the professional to proceed, get an estimate of the likely costs. If you so choose, put an initial ceiling figure on the amount you want to pay and get the professional to check with you before exceeding that figure.

Conducting a patent search with a searcher or attorney is a cheap deal, in terms of money, because it helps you to find out if you're wasting time pursuing your idea.

Using Your Search Results

After you have your first patent search results, what do you do next? You found some prior patent specifications showing inventions with similarities to yours. Look for the useful differences and distinctions that your invention offers. These features provide your innovation with its novelty and inventive step (see the earlier section 'Determining Whether Your Idea Is Really New' for more on innovative features) to make it patentable – and support its market value.

If you find prior art exactly the same or very similar to your invention, with nothing in your ideas that improves on what's known, think hard about whether you're ready to hit the marketplace. Those prior inventors went through the same steps that you're now going through. Ask yourself, 'Is my product different enough not only to be patentable but also to be commercially viable?'

If an exactly similar patent is still in force then get a legal opinion as to whether or not you may infringe it. If you're advised that you probably would infringe the other patent, stay at home – don't risk spending your hard-earned money in an expensive legal action.

If you're satisfied you wouldn't infringe, you may still want to contact the inventors of similar patents and find out whether they marketed their invention. And if not, why not? Did a market exist for their product? What went wrong – were the manufacturing prices too high, for example? If a market does exist and the inventor simply isn't capable of marketing the product, you may want to consider a joint venture with that inventor or even consider licensing his invention from him and paying him a royalty. (See Chapter 22 on licensing.) You have many options to consider.

In some cases, a product like yours may be on the market without a patent. Unpatented products can still be marketable, and millions of such products exist. However, if you choose to market your product without a patent you lose the right to patent it. Patent offices see no point in rewarding you for revealing an invention to them that you have yourself disclosed to the public. When a successful product with no patent protection comes to the notice of competitors they can produce their own version, and share or take over your market – and generally there's little you can do about it.

Looking pretty

If your product's appeal lies in its visual appearance and your searches reveal nothing visually similar, then remember the protection available for designs. You automatically have a design right from the moment you create a three-dimensional design and can use that to restrain competitors. You can gain even stronger rights by applying to register the design, whether two- or three-dimensional. See Chapter 3 for the basics of design right and registered designs, Chapter 7 for procedures to register a design, and Chapter 9 for the copyright background to design right and how to safeguard it.

Considering trade marks

Don't forget possible trade mark rights. If the searches show that someone else already owns your proposed trade mark in a classification similar to your product, it's best have a rethink and choose (and check out) another one. The existing owner is unlikely to want to license the trade mark to you, even with safeguards on product quality, and do you actually want to offer your product under a brand name that's associated with someone else?

If the searches show no registered trade marks like yours, think seriously about filing your own trade mark application. Unlike patents, you can still apply to register a trade mark after you use the mark in public, and you can renew the trade mark registration indefinitely, subject to paying renewal fees (in the UK every ten years). Check out Chapter 8 for more on how to register trade marks and the fees and other costs involved.

If the product is patentable and the trade mark is capable of being registered, we recommend that you get your product, and especially the trade mark, established in the market while the patent rights are still in force. Your best move is to register the trade mark at an early stage and work towards establishing a strong brand following that can outlive the patent.

Part II
Protecting Your Idea

'A bit drastic for a fly-killer, don't you think, Mr Prongdew?'

In this part . . .

The chapters in this part tell you how to use patents and other kinds of intellectual property rights to protect your invention. They show you how to apply for a patent, and how to protect it after you have it.

This part also guides you to other intellectual property rights: Some that come your way automatically and some you have to apply for. The chapters describe automatic design rights and other forms of copyright, how to register your designs and trade marks officially, and how to take care of them, too.

Chapter 5

Applying for Your Patent

*N*ecessity may be the mother of invention, but patents are the guardians.

This chapter shows you how to prepare a patent application that impresses everybody who sees it: the patent examiners who decide the scope of protection they grant you; the investors who may want to back your project; the manufacturers who may want to license it; the potential infringers who (we hope) back off when they see the quality and strength of the granted patent, and – if you ever end up there – the judge(s) in court who decide whether the patent is valid or whether it covers activities that infringe your rights.

Throughout this chapter we also try to alert you to formality traps that can destroy your patent rights, even if your invention is much better than sliced bread: things such as paying the official fees a day too late, paying the wrong fees by using an outdated fee schedule, or getting your co-inventor's name wrong. You can remedy some of these issues – if you spot them in time – but others you can't solve, especially in matters of fee payments. Losing your rights on such a technicality isn't a smart move. So read on to find how to steer clear of such problems.

Knowing Who Can Apply for a Patent

The people who can apply for a patent on your invention include:

▶ You, the inventor, along with any *co-inventors*, or joint inventors.

▶ Your employer, if the invention is made in the normal course of your duties.

    ✔ Your *assignee*, who is the person or company you sell your rights to.

    ✔ Your *patent attorney*, who is your guide along a path full of pitfalls.

This section looks at the technicalities of applying for a patent with other people or choosing someone to present the application for you.

Take care when you name the inventor(s). Patent offices can revoke a patent granted to a person who wasn't entitled to it. And if someone feels he is a co-inventor and you don't name him as such, he can apply to be added to the patent.

Applying with others

Many inventions have two or more people as co-inventors. Their patent application must be made jointly, or at least name each one individually, as an inventor. In the absence of an assignment or other such agreement transferring ownership of the invention to the intended owner(s), each joint inventor has an equal share in the patent and – if they so choose – can exploit it without reference to the other inventor. You can imagine the disputes this situation can cause. Make sure that you agree with any co-inventors the terms of your joint rights in writing – and if at all possible before you file the patent application.

Only people who actually contribute features of the invention are co-inventors. A person who merely provides advice or other assistance – such as where to get a special lubricant for the device's gearbox – isn't an inventor. Nor are financial contributors, which means anyone funding the research and/or development but not undertaking any of the work in reaching the invention. Listing the name of a financial supporter as a co-inventor is a nice gesture for maintaining his support, but not a good one legally! Similar thoughts arise in relation to your spouse. Sure, they've had to endure all those hours while you're away in the shed but you can find other ways to thank them than calling them co-inventors, if they are not.

One good way of involving non-inventor financial backers (or your spouse) in your patent is to include them as co-applicants. But make sure that you settle and record in writing the respective ownership and other rights of each of the co-applicants (even for your spouse). Another way is to set up a company and assign the patent rights to that company. Your co-applicants can have shares in your company and indeed you may want to open it to other shareholders too. You can get free advice on forming your company from such bodies as Business Link (see Appendix B for contact details) and from the later chapters of this book (especially Chapter 16).

Choosing a representative

If you think that applying for a patent by yourself is a way to save money, think again.

A good representative can help in drafting a patent specification that clearly defines your invention and shows its differences and improvements over anything known before. A good specification can help in getting the patent accepted by the examiners. A good representative plus a good specification can have a major bearing on the outcome of any negotiations or court proceedings. Choose well!

Although an experienced patent attorney doesn't come cheap, hiring someone qualified doesn't cost nearly as much as losing the rights to your invention. Especially if you're counting on your invention to provide a source of income, you need to make sure that your patent is as near perfect as possible.

Choosing an attorney you get along with, that you understand, and with whom you have a rapport is important. In the words of one of the UK Court of Appeal judges, Lord Justice Jacob, 'If you can't understand your patent adviser, get another one!'

If you simply don't have the money at first to pay for an attorney you can contemplate writing and filing your own patent application, although we repeat that this is not our recommended route. The filed application gives you an official receipt that confirms your possession of the invention at the date of filing. You're then able to approach third parties – would-be investors, manufacturers, or licensees – safe in the knowledge that you can show you invented it first.

Unless you're unusually proficient in technical writing and patent law, there's a significant risk that a self-prepared patent application may leave unprotected aspects that can let a competitor get around your patent claims.

To overcome at least some of the risks in filing your patent application unaided, a common solution is getting together with an interested (and asset-rich) third party soon after filing the application and reaching an agreement with them under which they take on the responsibility for strengthening and pursuing the patent application, possibly by preparing and filing follow-up applications. The deal should include the third party's payments to you, for example by one or more of a lump sum, of ongoing royalties, and ongoing technical consultancy.

Filing a Patent Application

The most common route, and indeed the recommended one, for starting on the road to patent protection is first to prepare and file an application at the official intellectual property office in your home country. This section deals mainly with the requirements and procedures of filing a patent application at the UK Intellectual Property Office (UK-IPO). Filing in Ireland and other countries in Europe involves generally similar activities.

You don't need at this stage to do anything about filing equivalent applications to cover other countries. You have 12 months from the home filing date before they need be filed. Do however be mindful of where you may in due course want to file equivalent applications.

The patent specification

Preparing the patent specification requires much thought and care and probably quite a lot of time. The following discussion on its contents is intended to help you understand what needs to be included, whether by an attorney or by yourself. If you do employ an attorney, you can greatly assist – and keep down the attorney's charges – by providing for them a clear and concise written description of your invention. Try to be as broad as possible, indicating its many applications and selling advantages. You know your invention much better than the attorney does. A good framework is shown by the following typical contents headings often found in United States patent specifications:

1. Title
2. Field of the Invention
3. Background of the Invention
4. Description of the Prior Art
5. Summary of the Invention
6. Description of a Preferred Embodiment
7. Drawings (if any)
8. Brief Description of the Drawings (if any)
9. Examples (if any)
10. Claims
11. Abstract

Items 2 to 6 and 8 and 9 form what's generally known as the patent's *description*. Don't get cross with your attorney if the draft description sent to you for approval omits some of them. Different inventions require different treatment to describe and define their key features.

The specification should be typewritten in black ink on white A4 paper. Rules exist on the size of margins at the paper's top, bottom, and sides; some specifications require 2.0 cm, others 2.5 cm. A good rule of thumb is to leave a 2.5 cm margin all round.

It's important that the description fully explains the invention at the outset because it's difficult, if not impossible, to add information to it later.

Title

The title heads the first page of the description. It should give a short indication of the field of the invention (the technical or scientific area in which your invention falls). Along with the name of the applicant(s) and the official application number and filing date, this detail is the only item published in the first official records on the application (usually within a week or so of the filing date).

Choose a title that doesn't give away too much of your invention so soon after filing. The title 'Vacuum cleaner' is fine, even though it does of course alert Dyson and Hoover to the fact that you're looking into one of their market areas. But don't try to disguise the invention too much. A title that simply says 'Device' may provoke the formalities examiner to publish *as the title* much, or all, of your main claim. Now that eventuality really *could* give it all away before you're ready.

Field of the invention

The opening paragraph is a single sentence statement of the field of the invention: the technical or scientific area in which your invention falls. Say your invention relates to a door hinge for motor vehicles, your statement may read, 'This invention relates to motor vehicles, specifically to an improved vehicle door hinge.' Don't get into the fine detail of your invention yet.

Background of the invention and description of the prior art

These sections are much the same. The term *prior art* includes all the published information and knowledge in your field: patent specifications, other publications, or a known product. Here, you describe what people have done before in the field of your invention; you review the prior patent specifications that have made similar proposals. Importantly it lets you point out disadvantages of the prior proposals and how you've overcome them.

You may feel hesitant about telling the examiners of the *prior art*, especially if some of it is quite close to your improved version. But remember that examiners are pretty effective searchers and can probably find the close prior art anyway. You're more likely to persuade an examiner to grant you a patent if you clearly acknowledge the prior art and explain from the outset what is better about your version.

Some examiners get really crabby if you don't tell them about prior art you clearly knew about. This is especially the case in the USA, where failure to reveal all is regarded as fraud and is a ground for revoking the patent (understandably so if in effect you've obtained your patent under false pretences).

Summary of the invention

In the summary you go into a bit more detail about your invention and its advantages. Here's also a good place for a statement of invention with wording substantially the same as the main claim you're including or intending to include. You can also usefully include statements of invention on preferred features corresponding to your intended subsidiary claims. See below for more on the claims section.

Description of a preferred embodiment

You must include a fully detailed description of at least one preferred embodiment of the invention. Just stating, 'The invention relates to a motor vehicle device that warns a reversing driver of the presence of another vehicle or lamppost to the rear' isn't enough. You must provide a description of at least one way of achieving that desirable objective and the working elements and their functions to achieve it: a particular construction of radar gun; an electrical alarm circuit involving a broken light beam, or simply an angled mirror hanging on the tailgate. The length of your descriptive section depends on the nature and complexity of your invention but it must give enough instructions for a reader to put the invention into practice after the patent has expired.

Drawings

Not all inventions need drawings. Most chemical, biological, and information technology inventions don't. But if the subject matter of your invention necessitates drawings, you must include them with your application. In the printed versions most IP offices put them at the end of the specification, though the US Patent & Trademark Office (USPTO) puts them at the front.

The drawings to be published need to be of a quality that permits photocopying, as they form part of the published specification. Try to ensure that they look good. Give some thought to the views they need to display to show the invention to good effect. You can get away with informal versions initially and the patent examiners later let you replace the informal ones by good formal versions before publication.

Many third parties, notably competitors and potential infringers, make their first decisions on the strength of your case by looking at the drawings. If the drawings look scruffy and carelessly drawn, the third party may see you as an easy touch and decide to infringe anyway. Several excellent firms exist that are familiar with preparing patent drawings (Appendix B provides contact details). They offer great value for money and their drawings will impress you and everybody else.

Resist the temptation to include printed matter on the drawings. Label the drawing with just the sheet number (Sheet 2 of 4), the number of the figure (Figure 1), and the reference numeral and the lines (the lead lines) joining each number to its respective element. The invention can then be fully covered within the description by reference to the individual numerals.

Brief description of drawings

The description relating to the drawings should refer to the constituent elements in turn, by individual reference numerals, which need to be repeated with each mention of the element, for example 'the pump (25)'. The best approach may be to work first through the structure of the device, describing what each element is, and then to start again, this time describing how each element works.

Examples

Many device inventions don't need worked examples. But most chemical and biological inventions do. For example, chemical inventions may need to specify the reactants (and not just by their brand names), their proportions, when they need to be added, how they should be mixed and for how long, the temperature and pressure conditions for the reaction, and any preparative or clean-up steps. And ideally include several examples, not just one. Multiple results can conveniently be shown in tables.

Claims

The claims form the foundation and nucleus of your patent and define the scope of your patent protection. They're one of the most important features of your application. They specify what you believe to be the novel and inventive features of your invention. Whether a patent is granted is determined primarily by the patent claims. And they're the primary features fought over in infringement actions.

Encapsulate each claim in a single sentence, effectively the end of a sentence that starts with the (omitted) words 'We claim . . .'.

Many first-time inventors fall into a trap regarding their claims. They present as claims various statements of the great advantages their invention offers, for example 'My improved bicycle goes twice as fast as any other; it's much easier to pedal uphill; its gear train will last 100 years.' Conceivably, all these wondrous claims may be true. But they don't make a patent claim!

Think instead of the claims as a fence around your back garden – they define the limits of your property; somewhere other people need your permission to enter. You want your fence to be as broad and sturdy as possible. Patent attorneys try to make the first claim as broad as they can while still distinguishing it from what has gone before. But they also include second and probably several other claims, each narrower in a different respect from the main claim: building up a set of claims and so providing several possible lines of defence if the broader ones can't be sustained.

Let's say the invention relates to a bicycle wheel with special spokes configured to give the wheel a spring effect. The main claim may be directed to the spoke configuration in general terms. Successive subsidiary claims may refer to the preferred individual spoke shape; the preferred spoke dimensions; the preferred plastic, metal or alloy for the spokes; the preferred way at securing them to the wheel, etc. Each of these subsidiary claims may have novel and inventive features to support the grant of a patent to you, even if the general principle of spring-effect spokes defined in the main claim turns out to be already known and thus not patentable.

Many examiners, especially at the European Patent Office, insist upon the claims being in the 'two-part' form, in which the first part of the claim (the *preamble*) sets out what's already known and the second part (the *characterising part*) tells what's new about your invention. The two-part claim is however not always the best way of defining the invention. A number of good inventions are simply a new combination of known features put together in a way that produces surprising advantages.

Try to avoid unduly limiting the claims. If the prototype is made of stainless steel – because that best withstands the required duty – don't limit the claims to stainless steel. If you do, you make it more likely that your competitors can avoid your patent by using a different material such as aluminium or an engineering plastic. Spike their guns by using wording to include not only stainless steel but also all other reasonably suitable materials, which you should name in the description.

Abstract

Prepare the abstract on a separate sheet of paper with the heading 'Abstract'. It needs to include the same title as in the application form and on the front page of the description. The abstract is in due course included on the first page of the published specification. It should be a concise summary of your invention, as brief, complete, clear and to the point as possible. The abstract is generally the part of the application that's read first. You may want to view abstracts from other patents to give you an idea of what's needed.

If your specification includes drawings the published specification includes one of them alongside the published abstract. You need to nominate which one you would prefer it to be. And the abstract needs to include after each mentioned component the equivalent reference numeral from the drawing.

Many new products and processes involve more then one invention. Your innovative bicycle may not only have new and improved suspension but also a modified wheel structure and/or modified tyres to work with the new suspension, maybe even a tyre pump tidily hidden inside the frame. Big companies would file a separate application for each of the inventions. Lone inventors usually can't afford that. The lone inventor can however include all the inventions in a single initial application, so giving all the inventions the same priority date. The patent examiner may object to the presence of more than one invention and choose to search and examine only one of them. The others can however be 'divided out' of the first application at any stage up to its acceptance and so become one or more separate applications. Subject to payment of the fees for each such separate application, and provided that the 'divided out' invention is novel and inventive, it can go on to become a separate granted patent.

Paying attention to the process

An application for a patent must include the patent specification, with its full *description* of your invention, any necessary *drawings*, a set of *claims* defining your invention, and a short *abstract*, plus a completed *application form*, a *fee sheet,* and any necessary *fees*.

However, some of the elements of the application need not be filed with the original papers, as long as they follow within defined periods. The required elements at day one are the application form (UK Patents Form No. 1) and the full description (and any drawings). In the UK not even the application fee is essential at day one, though oddly it's advisable to file a fee sheet for zero pounds as doing so helps the UK-IPO formalities.

Similarly some other elements of the application can be included with the original papers, or filed later. These elements include the request for a preliminary search (UK Patents Form No. 9A) and a request for substantive examination (UK Patents Form No. 10). A statement of inventorship (UK Patents Form No. 7) is also required if the person filing the application is not the inventor or if the application is being made in the name of a company.

The set of claims can also wait – in principle – but until it's filed the official examiners can't conduct their searching or examination, because you haven't defined for them what you consider to be your invention.

The application can be filed by post to the UK-IPO (Concept House, Cardiff Road, Newport, NP10 8QQ) or online (the Web site address is given in Appendix B). You get an official filing receipt, usually within a few days of filing, showing the allocated application number and confirming the date of filing. You're now in a position to talk to third parties about your invention because your priority date has been formally recorded.

Provided that you've paid the official application fee, the UK-IPO makes a formality check to make sure that all essential parts of the application are there and that other necessary fees have been paid, and writes to draw your attention to any missing elements.

The UK-IPO adopts a very user-friendly approach to lone inventors and guides them, usually by letter, on any element missing from the application and the date by which it's required. But don't let this encourage you to be sloppy in what you present to them!

Publication

Around 18 months after the filing date of the application, and provided you've met all the formal requirements, the specification is published. Anyone can now buy a copy or inspect it online and read every word of it. What may sound like bad news can in fact be very good news. Anyone who chooses to infringe your published patent risks having to pay you damages from the date of publication, although you can't institute action to try for damages until the patent is granted. In the meantime you keep them guessing as to what the eventual scope of your claims is going to be. Such a risk is enough to make many potential infringers hold off until the patent is granted and they can see just what they must avoid.

You can request publication before the normal 18 months, for example if you become aware of someone infringing, or thinking of infringing, your patent. This brings forward the date from which damages can run and can be very important for products with a short market life. Think of the 'must have' items for Christmas – the market may not last for more than their first year.

On the other hand if you decide to avoid publication, and so keep the invention secret, you can withdraw the application before publication, though with the loss of the prospect of getting a patent.

Search

After you've requested a search your file goes to a patent examiner who is trained in the field of your invention. The examiner searches for prior inventions and proposals with similarities to yours: patent specifications from anywhere in the world or other published documents or information. A search report is then issued to you, usually within a few months, listing the findings and classifying them into those that may limit your scope of claims and those that give merely background interest.

Chilled out

Examiners can look far and wide for prior art. In an application for a label that would change colour to indicate the temperature of the product to which it was applied, the UK examiner cited a bottle of Newcastle Brown Ale (but for unexplained reasons didn't include a bottle, not even an empty one, with the official letter). The *brown ale* bottle had a triangle on the reverse label that changed colour to indicate when the recommended level of chilling had been reached.

Substantive examination

Substantive examination is so called to distinguish it from all the other examinations the application faces. This examination is the big one, the one where an examiner compares what you claim as your invention with whatever has gone before, in particular with what was found in the search stage. You would normally request substantive examination after the search report has been received and studied, because doing so allows the specification to be amended with a view to improving your prospect of being granted a patent. In the UK it is however possible – and often sought if quick progress to grant is required – to request search and substantive examinations at the same time.

Sometimes, though rarely, the substantive examiner raises no objections to the application and it proceeds directly from examination to the granting formalities. More often you'll receive an official letter from the examiner, which does at least one of the following:

🗸 Objects to one or more parts of your specification.

🗸 Rejects some or all your claims due to unclear language or lack of description in the specification.

🗸 Rejects some of all of your claims due to prior art.

In replying to the official letter you can offer corrections and amendments to the specification to deal with the objections, although you must not extend the scope of claims beyond the original filing. A first reply may be enough to satisfy the examiner but further exchanges of official letters, replies, and amendments are often required. A UK patent application typically takes two to three years from priority date to grant. In any event the application must be put in order for grant within a maximum of four and a half years from the priority date. If you can't satisfy the examiner within that time frame the application is void.

Equivalent Applications

In today's global market you're likely to consider protection in other countries of interest. This section describes the various ways in which you can ensure protection abroad.

For the purposes of this section we assume that you've first filed an application at the intellectual property office in your home country (which we assume to be the UK-IPO). Filing outside the home country in the first instance is possible but, primarily for security reasons, your home IPO may require you to get their prior consent. Check with them first.

National applications

Provided you file an equivalent national application in another country within 12 months of the priority date of your home application, that further application retains the home priority date (a useful aspect of the Paris Convention – see Chapter 3). Each such national application requires local application forms and fees on filing and progresses through similar stages of publication, search, and examination through, if successful, to grant in that country.

There's little prospect of filing and prosecuting the application yourself in the other countries: Almost invariably local attorneys need to be used. To ensure the use of local attorneys who can offer the expertise required for your invention the safest route is to instruct attorneys in your home country and let them prepare and send documents, in a format they know meets the local requirements, to associate attorneys they know to be capable and trustworthy.

You're normally asked to sign a *Power of Attorney* document to authorise the associate attorneys to file and manage the application in their country. The wording – check it if you feel uneasy about it – is normally limited to activities in the immediate application. Some big firms give a general power of attorney to the associate attorney to act in all their IP matters in that country, for the convenience of all parties, but this would not normally be the course for a lone inventor.

A huge drawback of filing individual national applications is that each carries its own set of official fees and professional charges. And because of the large number of different languages around the world, you normally find only a few in your filing programme that use your native language. So you're often faced with substantial translation costs at a fairly early stage in your invention's life. And good quality translations don't come cheap.

International applications

If your overseas interests extend to two or more countries it's almost certain to be most cost effective to look to one of the international application systems under which at least some of the procedural steps towards a granted patent can be conducted in a single application. For international (PCT) and European (EPC) applications from UK-based applicants the initial filings may be made, in English, at the UK-IPO who forward the papers to the appropriate organisation for further processing, again in English. Translation of the specification, or at least of the granted claims, is eventually needed in non-English speaking countries where the patents are granted or validated but the costs of the translations are put off for several years.

The Patent Cooperation Treaty (PCT)

If your countries of interest extend beyond Europe the easiest and cheapest route is likely to be an international (PCT) application. You can designate any or all of the member states (137 in 2007 and still growing). If the PCT application is based on an application in your home country it must be filed within the 12-month Paris convention period to retain your home country priority date.

Although the body responsible for PCT applications is the World Intellectual Property Office (WIPO) in Geneva the search and any examination stages for applicants based in the UK and other member states of the European Patent Convention (EPC) are undertaken for WIPO by the European Patent Office.

The PCT application must include the request form (PCT/RO/101), the official fees, and the PCT specification with contents as in the specification in the priority application in your home country – with possible additions at this PCT filing stage, although any such additions won't have the home country priority date. Online filing is also an option (see the WIPO Web site: www.wipo. int/pct/en). You can designate the member states you require, either as individual countries or by regions, or by both of these options. Four systems allow regional designations: the European Patent Convention (EPC, covering 32 countries in Europe), the Eurasian Patent Organisation (9 former Soviet countries), the African Regional Intellectual Property Office (ARIPO – 16 countries, mostly in East Africa) and OAPI (Organisation Africaine de la Propriété Industrielle – 16 countries, mostly in French-speaking Africa).

The most notable PCT absentee is Taiwan: if you want to file there you need to file a national application (and send the papers in good time for the translation to be ready within the 12-month convention period).

The PCT steps are substantially the same as the equivalent steps in most national applications: filing receipt with official number and filing date; publication at 18 months from priority date (or faster on request), and a prior art search and report. Optionally (and with a further fee) you can request preliminary examination, which leads to an official international preliminary opinion on the patentability of the application. The PCT opinion is sent on by the EPO to all designated regions and countries.

The PCT preliminary examination system is very user friendly. The examiner first sends a written opinion to the applicant (you or your attorney) indicating aspects that require attention. Effectively this tells you the amendments necessary to put the application into an allowable form. You or your attorney submits amendments to correct the deficiencies, making very likely the issue of a positive international preliminary opinion.

The PCT system does not however directly grant the patent. Within 31 months from the priority date (for UK applicants) your PCT application must be taken into the regional or national phase, meaning that your route to gaining patent protection now continues through your chosen regional or national IP offices.

Using one or more of the regional systems is not essential but, as with the PCT itself, they do offer significant cost savings compared with separate national applications, Beyond the 'regional' countries, for example for the USA, Japan and South Korea, the application must proceed as an individual national application.

The European Patent Convention (EPC)

The most commonly used regional system for UK-originating applications is the EPC. The EPC application can be filed, in English (or in French or German), at your home IP office (in the UK at the UK-IPO) designating any or all the member countries. If you've taken the PCT route you can enter into the EPC phase from there.

The EPC members include all the members of the EU, plus Iceland, Monaco, Switzerland & Liechtenstein, and Turkey (but not Norway, although it's a member of the PCT). A granted EPC patent can also be validated in one or more of the so-called extension states, currently (2007): Albania, Bosnia/Herzegovina, Croatia, Serbia, and the Former Yugoslav Republic of Macedonia.

The EPC application is processed by the European Patent Office (EPO) as a single application. The processing goes further than that of the PCT. After publication and a prior art search (which may have been conducted at the PCT stage) the application undergoes a full examination. If accepted it becomes a granted European patent which can be validated as separate patents in your designated countries. Enforcement of the patents – if enforcement ever becomes needed – is undertaken by proceedings in the respective country's courts.

Because the UK is a PCT member you can choose to pursue UK cover via the PCT and abandon an original UK application for all purposes except for proving the priority right. Some applicants however choose to continue with both the PCT and UK applications, giving them maximum flexibility in choice of route as the applications proceed.

If you do continue with both UK national and EPC applications for the same invention you must eventually choose between being granted a UK patent as such or an EPC patent validated and effective in the UK. You can't patent the same invention twice for the same territory!

Considering Costs

Applying for and getting a patent is not cheap! Not only must you pay official fees at many stages, some of them substantial, but if you use attorneys they usually add a handling charge for dealing with the action the fee involves – and that handling charge is often at least as much as the fee itself. When it comes to the professional services of attorneys in preparing patent specifications and dealing with examiners' objections, their charge rates reflect their high levels of skills and the substantial administrative costs their office has to bear in reliable diary and reminder systems, professional insurance, and experienced support staff.

When reflecting on the high costs of getting patent protection, don't overlook the returns they can generate. In June 2007 New Zealander Juliette Harrington sold to Yahoo! her United States patent on one-stop Internet shopping for NZ$6.55 million (£2.25 million).

The following cost figures (based on 2007 fee schedules) are merely indicative. Official fees are reviewed from time to time, usually on an upward-only basis, and professional advisers' fees have a way of creeping upwards too. Also the amount of professional work involved can vary greatly from case to case. The figures quoted below assume a specification with about 4,000 words. The current official fees can be checked on the respective patent office Web sites (see Appendix B), and you can check in advance with your professional advisers as to how much their services for a particular activity are likely to be.

The figures below assume communications by surface and airmail. In many cases official fee savings can be made by filing online.

National applications in English

The charges for preparing and filing a patent application in UK or other English speaking country, using professional services and including attorneys' and official fees, are likely to be at least £2,000.

Taking such a national application through search and examination to grant is likely to incur a similar sum of at least £2,000 in total. However, you have no obligation to continue with the application if it runs into undue examination problems or if the market no longer justifies it.

National applications not in English

Translation charges add dramatically to the costs in applications using languages other than English. For European languages: French, German, Spanish, and so on, translation charges can increase the application charges to at least £3,000. For applications in Asiatic languages: Japanese, Mandarin, or Korean expect application charges of at least £4,000. Provided the local agent communicates with you in English (which is usually the case) the further charges through to grant may be similar to those for UK, that is, at least £2,000. Again, at any time you have the choice of whether or not to continue with the application and associated expense.

PCT applications

PCT applications carry substantial official fees of about £1,800 at the filing stage. The attorneys' charges can be at least as much again. The fee for the optional subsequent preliminary examination is about £1,100, plus the attorneys' handling charge. Attorneys' reporting on the search and preliminary examination, and submitting observations on the preliminary examination, typically adds at least a further £1,000.

EPC applications

EPC official application fees, assuming all member states are designated, are about £1,200. The attorneys' charges are likely to be at least £1,800. The official fee for the subsequent examination is about £1,000, plus the attorneys' handling charge. Attorneys' reporting on the search and examination, and dealing with the examiners' objections, are likely to add at least a further £2,000.

Validating your granted EPC patent in each of your chosen countries involves relatively simple formal steps, including translation of the claims for countries with a language other than that of the application. The charges per country are unlikely to be more than a few hundred pounds.

Renewal fees must be paid in each country in which you want to keep the granted patents in force (see Chapter 6).

Getting Your Patent

After months (perhaps years) of waiting, you finally receive your patent. You breathe a huge sigh of relief and feel the urge to sit back and relax. Not so fast! Although getting a patent is a very big step, it's still only another step in what can be a very long journey to making money from your invention.

Even if you get your product produced, your invention isn't going to make any money for you until you get commercial returns when your product starts selling in good numbers or you strike a good licensing deal. Even famous inventors hit clinkers all the time. Think of Clive Sinclair's C5. On the other hand, an idea so seemingly stupid that you'd think that it was somebody's idea of a silly joke can do really well.

Face it; all a patent does is give you a right to prevent other people from making, using, or selling your patented product. You still have to produce it, market it, and sell it. Look to the chapters in Part IV for advice on how to commercialise your invention.

You must be vigilant about maintaining and protecting your granted patent. It's your responsibility and no one else's to do so. The IP offices issue the patents but it's the patent owner's job to enforce them. See Chapter 6 on how to maintain and defend your patent.

Chapter 6

Maintaining and Defending Your Patent

. . . .

In This Chapter

▶ Keeping your patent up to date

▶ Placing a value on your patent

▶ Sharing your patent rights

▶ Protecting your property

▶ Discovering patent insurance

. . . .

*J*ust like your home, a patent is a real piece of property. You have to tend to its upkeep and protect it from squatters and burglars. And just as the bank that holds your mortgage doesn't pay for your home insurance, the intellectual property (IP) offices don't enforce your patent. You must enforce it yourself, or hire someone who can.

This chapter tells you how to keep your patent current and how to keep it yours by defending it if need be.

Keeping Current by Paying Your Fees

Don't think that the official fees you pay with your application and for search and examination are all you have to pay the IP office. After the IP office grants your patent – and in some cases while the patent's still pending – you must pay renewal fees (also known as maintenance fees) in order to keep your rights alive. The following sections offer advice on how to avoid missing renewal dates, and consider the likely cost of renewal fees.

Remembering renewal dates

In most countries and regions you pay the renewal fees annually on the anniversary of the filing date. The current renewal needs are as follows:

- ✔ **United Kingdom Intellectual Property Office (UK-IPO).** In the UK the first renewal (for the fifth year) potentially becomes due on the fourth anniversary of the filing date. However, the UK-IPO gives you four and a half years from filing to get the application into shape, so if your UK application is still pending within this time period then the UK-IPO defers the date for the first renewal fee.

- ✔ **European Patent Office (EPO).** The EPO aren't quite as kind as regards pending applications. Their first renewal fee (for the third year) becomes due on the second anniversary of the EPC filing date.

- ✔ **United States Patent and Trademark Office (USPTO).** A different renewals regime operates in the USA. The USPTO require renewal fees for a patent (but not design or plant patents) at 3½, 7½, and 11½ years from the date it grants the patent.

Facing up to the costs

IP offices increase patent renewal fees on a sliding scale over time (and tend to revise fee schedules upwardly every year or so). The following figures are taken from schedules and exchange rates for 2007:

- ✔ **UK-IPO.** Fees start at ₤50 for the fifth year, increasing annually to reach ₤400 for the twentieth year.

- ✔ **European Patent Office (EPO).** Renewals start at €400 (₤270) for the third year, increasing annually to a maximum of approximately €1,065 (₤725) for the tenth and each subsequent year. (Fortunately most applications at the EPO reach the grant stage within a few years.)

- ✔ **USPTO:** Renewal fees in the US, like other US patent fees, depend on whether or not the applicant or patentee is a *small entity* (an individual or a company with less than 500 employees). For a small entity the fee is $450 (₤225) at 3½ years, $1,150 (₤575) at 7½ years, and $1,900 (₤950) at 11½ years. For companies other than small entities the USPTO doubles the fees.

The IP offices don't send you a reminder or invoice when your renewal fees are due. If you don't pay your renewal fee on time, or within a permitted extension period (typically six months and with a penalty charge for late payment) your patent rights may lapse, making your invention up for grabs for anyone to use without paying you anything. If the patent lapses you may still be able to restore it, subject to a restoration fee and adequate reasons for why you missed the payment (perhaps illness or loss of key staff). But even if the restoration is successful, you won't be able to sue a third party for operating within the scope of the patent during the period it was not in force.

To avoid late renewal payments, with the penalties and extra work they involve, keep good diary records, or pay an attorney or renewals agency to alert you in good time that the payments are due.

If you employ an attorney or renewals agency to handle your patent renewals, remember that you're merely transferring to them the heavy responsibility of keeping you alert to forthcoming renewals and making sure that you pay these in the right amount and at the right time. So each renewals payment carries a handling charge, and in the early years the handling charge is likely to be more than the fee itself – and the money to pay for the diary system, reminders, and payment activity has to come from you! The increasing size of renewal fees over time means that they can get quite expensive for the independent inventor. The IP office thinking is to encourage you to release the invention to the general public – by not paying the renewal fee – if the costs of renewal outweigh the returns you get or are likely to get. So if your invention isn't making money, consider letting your patent lapse.

You can reduce your UK patent renewal fees by applying to the IP office (on Patent form No. 28) to enter into their official records that licences under the patent are available to any third party as of right. If you and a third party can't agree on the payment or other terms of the licence either of you can apply to the IP office (on Patent form No. 2) to set the terms. The renewal fees for a patent carrying a 'licence of right' entry are reduced to one half of the scheduled figures.

Valuing Your Patent

Patents and other forms of intellectual property, like any other kind of property, have intrinsic value. Certainly, your patent is valuable to you, mainly because you hope it will make your fortune. So if you pay renewal fees (see the preceding section) or defend your patent in a courtroom, you want to be sure that you can justify the money you have to lay out.

The following sections look at placing a value on your patent and methods of sharing the expense of maintaining a patent.

Taking a businesslike view

Businesses often compare the value of intellectual property to that of the goodwill a company builds up with its customers and suppliers throughout its operation. Putting a money value on a patent can be difficult, but is actually easier than figuring out how much a customer's good opinion of a business is worth.

To evaluate a patent's worth, look at actual sales numbers if the product is on the market. If your product's not on the market, estimate market share, make projections of the potential market share you plan to capture through growth, consider the competition, look at the overall industry, and translate that information into monetary values. If your patent covers a process, look at the savings it makes compared with other processes that meet the same purpose. See Chapter 18 for more on market research and estimating your costs and profit margins.

Another factor in assessing the monetary value of the patent is its legal strength. How well can you defend the patent against attacks by third parties; how well did you write the specification and in particular the claims? We cover these aspects of the patent specification in Chapter 5.

Assigning your patent

If you work for a company whose business involves developing new products and processes, your employment contract or agreement probably requires that you to transfer (assign) to the company in writing the ownership of any patents arising from your work with the company. In return the company pays you to do the development work, whether or not the company can patent the results. You confirm the transfer by signing an *assignment*: a legal document that the law recognises as a transfer of ownership. Separate assignment documents may be needed for each country in which the company applies for a patent on your work. (See Appendix A for a specimen assignment document.)

As the assigning inventor, your name goes on the patent, but your patent rights go to the company. Sometimes, though rarely, inventor–employer agreements involve sharing the patent rights. Sharing arrangements are more common in situations with financial backers who want a share of the patent rights in return for their money. Chapter 11 has more about assigning rights.

Licensing your patent

Most inventors don't have the skills or the time to undertake their own production, marketing, sales, and distribution. Instead, they try to license their patents to companies who do have the necessary resources to make inventions profitable. In a licensing agreement, you give permission to someone else (a company or an individual) to manufacture, market, or distribute your product in return for a set amount of money or for a *royalty* – a percentage of the proceeds. Chapter 21 explores licensing in more depth.

Defending Your Patent Against Infringement

If your patented product proves to be a great success in the market, it may only be a matter of time before you get to hear that unauthorised versions are on offer. Someone is infringing your rights. What can you do about it? The following sections take you through dealing with infringement.

Checking your facts on infringement

First, check out the facts. Compare the rival product with the wording of your claims: They're the means by which a judge decides the scope of your protection and whether it covers the rival product. If the product doesn't use certain key features of your claim, it may not infringe your patent, though if you have a good invention and well-constructed claims, the missing features should mean that it's an inferior product to yours. If the product does appear to fall slap bang in your claimed area then you may indeed be facing an infringer. If the product somehow manages to avoid falling within the strict wording of your claims but still looks very close to them, a judge may still hold the product as an infringement. Judges generally don't like people who manage to skirt around patent claims while taking the benefit of the invention.

 Your rights against a possible infringer date back to the publication of your patent specification (which occurs 18 months or less after the filing date of your first patent application for the work). Although you can't bring the infringement action until you have a patent, the infringer may have to pay you any damages eventually awarded with effect from the publication date.

If you're satisfied that you may be facing an infringer, seek professional help. If you used a patent attorney in preparing your patent application and taking it through the official procedures, call them again. If you prepared the application unaided you've reached a point at which specialist advice from a patent attorney or an IP solicitor is pretty much essential. Chapter 3 has more on seeking professional helpers.

Avoiding the litigation lottery

Your first thoughts may be to get your IP solicitor to start legal proceedings for infringement against the retailer, importer, or stockist who is marketing or controlling the offending product. Wait a minute! Doing so risks a potentially long, exhausting, and very, very expensive ride though the law courts.

Of course, you may end up in court to defend and enforce your patent. IP litigation (taking legal action) is becoming increasingly common, especially among high-tech products and companies. However, you can try several avenues before reaching that point, but do take legal advice on which best suits the position you're in.

One of your first problems may be to identify the *defendants* (the parties offering the rival product). If the defendant is a large company selling through high street stores, locating it may be easy. Likewise, you may be able to identify an importing company. A bigger problem arises with street market traders, some of whom aren't averse to selling pirated goods and trying to keep beyond the arm of the law. For dealing with the latter group, you may get help from the local trading standards officers (contact them via your local town hall www.tsi.org.uk); they're none too keen on pirated goods and their attendant health and safety risks.

A first step may be simply to make the defendants aware that your patent exists. It's conceivable (just) that they don't know about your patent or assume that you'll do nothing to enforce it. But a trap exists here. If you don't choose your words carefully, you may find yourself on the wrong end of a legal action for making unjustified threats. Tempting though it may be, don't ever say anything along the lines of, 'Here's a copy of my patent. You're infringing it and I'm going to sue you for all you're worth.'

In the UK one way of alerting the defendants to your patent in a way that a judge won't consider to be threatening is to serve them with a formal claim (writ) for infringement. You IP solicitor serves this claim, which isn't a threat of legal action because you've actually taken the action! Just don't get carried away with the legal proceedings, at least not yet. Let that claim rest dormant while you pursue a settlement.

In some cases a meeting with the defendants, conducted with or by the respective legal advisers, may be enough to reach an arrangement. The defendants say sorry and agree to stop (well you can live in hope) or, perhaps more likely, agree to take a royalty-paying licence. Work out in advance a figure at which you're prepared to settle.

Choosing alternative dispute resolution

If early agreement proves impossible, consider trying *alternative dispute resolution* (ADR) – the collective term for ways that parties can settle civil disputes with the help of an independent third party and without the need for a formal court hearing, for example mediation or arbitration. You can choose between a number of different agencies to mediate or arbitrate between you and the defendant (see Appendix B for a list of some of these agencies). Your IP advisers can guide you as to which agency to choose.

Many people prefer mediation to arbitration as you're not bound to accept the mediator's recommendation. In arbitration, however, you agree in advance to accept the arbitrator's decision, even if you don't like it.

In a typical mediation the mediator first meets separately with each side and then together to discuss the issues in dispute. The mediator does not impose a decision on a solution but rather acts as a facilitator to help the parties find a solution to which they can both agree. The discussions aren't binding. If the mediation fails, the parties can continue with legal proceedings.

If a party in legal proceedings unreasonably refuses mediation or any other form of ADR, the hearing officer or judge may take such a refusal into account when awarding the costs of the action.

The UK-IPO actively encourages parties to consider mediation as a way of settling their disputes and assists in the selection of a mediator. If you feel uncertain about the strength of your position in the UK you can seek an independent assessment of the main issues in dispute from the UK-IPO. This assessment is designed to help you and the other side focus on the main issues in dispute and test the strength of your arguments.

Going to court

If all attempts to reach an out of court settlement fail and you find yourself in court, try to find the funds to get good representation (and let's hope that by now your product is selling by the cartload and you have a wealthy backer in tow). Your patent attorney or IP solicitor can help you to select the members of your presenting team. Depending on the country and court, you may need to appoint a specialist barrister or other advocate authorised to appear in the court in question.

Almost as important as good representation in court is that you behaved properly in your patenting, marketing, and business activity in general. According to Lord Justice Jacob, 'The good guy wins.' You want to show – as diplomatically as possible – that you're the good guy and the other side are unspeakable villains. Interestingly, your possibly small company and obvious lack of resources may help you: Courts tend to have a measure of sympathy for what they see as big business crushing the underdog.

Generally, in preparing for the court proceedings you first ask for an *injunction* – a court order to stop the infringement. You may also sue for the cost of damages due to the infringement (and these can go back to the date the IP office published your patent) and for the delivery of the infringing articles to you.

Probably the first thing the defendant does is question the validity of your patent. The judge makes certain factual enquiries to decide if questioning your patent validity is fair. This is where the money you spent on a good patent attorney and your choice of good trial lawyers really pays off.

In addition to being very expensive, court proceedings can last a very long time. And even if you emerge with a favourable decision from the first court, the defendant can appeal the decision to at least one higher level of court.

In the UK, the first level of court you go to may be the Patents County Court – in principle the quickest and least expensive option. Higher levels are the High Court, then the Court of Appeal, and in certain cases the House of Lords. And in theory you have to fight similar actions in every other country where infringement occurs. In practice, a successful action in one country may be enough to persuade defendants in another to settle.

Top Quality Infringement

Ron Hickman, the inventor of the Workmate workbench, won a number of infringement actions after the product became a runaway success. Part of the settlement in Japan was the delivery of the infringing versions to him. We heard that the infringing articles proved to be so well made that his company simply rebranded them, repackaged them, and put them back on the market.

The best way to avoid the costs, anguish, staff, and management time, and uncertainty of outcome of proceedings in the courts is never to go there! Do your utmost to reach a settlement before you get to the court door.

Insuring Your Patent

Defending your patent in court is costly. Patent litigation through trial probably costs at least £500,000 and can run into several million pounds, especially if the action goes to appeal. Then you have the hidden costs of litigation. If you're in the courtroom, you're not attending to business, so risking a variety of business problems. You and your family come under unaccustomed stress. You may also be in the centre of a large or small media storm, which just adds to the stress for all concerned.

In a patent infringement case, a bigger, wealthier company may overwhelm an independent inventor or smaller start-up company. For this reason, you may consider taking out patent insurance to protect you, the patent holder, against any losses you may suffer as a result of someone infringing your patent.

Like everything else to do with patent litigation, patent insurance isn't cheap. For cover up to £500,000, the lowest likely cost of an action, the premiums may be at least £5,000 *per year*. Before issuing a policy to you, the insurance company may seek a professional legal opinion from an independent patent attorney regarding the overall validity of your patent and related issues such as your company's competitors, the market for your product, and current patent cases and products in that market.

Whether to invest in patent insurance has to be a decision for you. It may give you a degree of reassurance and help you sleep more easily at night. But you can be fairly sure that, even with pretty hefty premiums, it won't cover all the costs of a prolonged or difficult action.

Don't let insurance coverage give you a false sense of security and lead you to litigate just because you can. Going to court may bring you negative publicity, and don't forget that you can lose!

Chapter 7

Applying to Register Your Design

· ·

· ·

*I*n contrast to a patent (which covers the way an article functions or the process of making it), a design relates to the appearance of the article. Think of design as the feature or features that 'appeal to the eye', though in intellectual property (IP) terms artistic qualities aren't essential: A distinctive shape with no aesthetic appeal may still count.

A *registered* design gives you legal rights and has some similarities to a patent, both in the procedures for applying for one and in the steps you may have to take to enforce the design. This chapter tells you how to set about applying to register your design.

Deciding Whether to Register

Design protection covers the outward appearance of your product, including decoration, lines, contours, colours, shape, texture, and materials. Registered designs can protect three-dimensional or two-dimensional features: shape, configuration, pattern, or ornament.

You acquire some automatic free protection upon creating the work in question. The protection includes an *unregistered design right* that protects the work's functional and aesthetic aspects and three-dimensional shape (but not its surface patterns, ornamentation, or decoration). Unregistered design rights may run for up to fifteen years in the UK and three years across the EU.

You also acquire *copyright* that protects the two-dimensional drawings or plans of the design, and the design itself if it's artistic and you don't intend to mass-produce it, for example that watercolour you painted for Grandma's birthday. (See Chapter 3 for more on unregistered design right and Chapter 9 for more on copyright.)

This section helps you decide whether applying for design registration is worthwhile and offers advice on choosing a professional to submit your application.

Considering the advantages

Registering your design offers certain advantages over the unregistered options. Registration in the UK and across the EU gives you up to 25 years' protection (subject to renewal fees – see the 'Considering Costs' section, later in this chapter). Plus, having the registration means that in the event of a dispute with third parties you avoid most of the lengthy and expensive stages of having to prove such aspects as the date of creation and your ownership of the design. You have an official design registration number and confirmed registration date that you can put to third parties or the courts with copies of the design illustrations to show your established rights.

To help you decide if design registration may be the route for you, ask yourself the following questions:

- ✔ Does automatic design right or copyright give the design sufficient protection without incurring the expense of registration?

- ✔ Has somebody else already registered the proposed design? (Chapter 4 explains how to check this out.)

- ✔ Which countries do you want to cover by design registration? (See the later section 'Applying to Register a Design' for filing options.)

To apply for a registered design you must be the rightful owner of the design – as its creator, or as a result of the creator assigning (transferring) the design rights to you. If you commission someone to produce a design and pay an agreed amount to them to cover the work involved then you're the owner. However, in order to avoid future disputes, arrange for the design's creator to sign a written assignment document, which confirms you as the owner, that the creator undertook the work for good and adequate remuneration and that the creator is willing to support applications for design registration. We include a specimen assignment document in Appendix A.

Unlike the situation with most patent applications, revealing your design before making the application need not be fatal. Some IP offices, including those of the UK and EU, allow a grace period of up to 12 months to test or try out your design before applying for registration. During the grace period you have a certain amount of protection from unregistered design right or copyright as discussed above.

Protecting spare parts

The issue of design protection for spare parts is a source of much dispute. On the one hand the manufacturers of the original article usually want to be the sole supplier of the spares. They're concerned that the spares are of good quality, and won't damage other parts of the product or harm their good reputation. They may also have design rights in the parts – whether unregistered or registered. On the other hand some spare parts supply companies that make quality replacements argue that many spares have little or no exclusive design content and that the original manufacturers use doubtful rights to sustain a monopoly in routine items that should be more widely available.

The UK IP office and courts employ principles known as 'must fit' and 'must match' to decide whether or not to give spare parts design protection. Broadly speaking, if the need to fit or match a given location dictates the shape of the product, there is no design protection. However, if the part contains design features that give it a distinctive appearance – and in the case of an unregistered design right these must be three-dimensional features – then it may warrant design rights.

Take motor vehicle spare parts as an example. You don't get design protection for exhaust pipework, because it must fit the available path to the rear from the vehicle engine; nor to body panels and doors, because these must match the shape of the rest of the vehicle. But you may get protection for wing mirrors, because they come in a variety of shapes not dictated by the need to fit or match; or for car wheel designs: A walk around any car park shows the huge range of designs of wheel shapes.

Choosing a representative

Using a professional representative to handle your design application may or may not be essential, depending on the country or region you want to cover. However, we strongly recommend employing a representative to help you register your design, because the laws, rules, and time limits can create unexpected difficulties that may prove fatal to the success of the application.

Most UK patent attorneys and trade mark attorneys have experience in design matters. The attorney is also in a good position to advise on whether design registration is a necessary or desirable part of your IP portfolio, and when it comes to protection beyond the EU, can select and instruct known and trusted associates to handle the local applications. And very importantly, the attorney can advise you on putting together a professional application that helps to convince competitors to respect your design rights.

Residents of UK or other EU countries don't have to use a professional representative when applying for an EU-registered community design. But if you do, the EU office that handles registered design applications (OHIM) requires that the representative is qualified in IP matters in an EU country and is on its list of authorised representatives.

Applying to Register a Design

This section takes you first through the procedures to obtain design registration through a UK national application covering just the UK. However, unless your market interests are confined to the UK, you may want to consider applying to register your design more widely. So the section next describes the option of applying for a single registration covering all the EU member states, including the UK. And then it describes seeking protection in countries abroad through national applications, country by country.

Applying for UK national registration

To register a design in UK you apply to the Designs Registry of the UK Intellectual Property Office (UK-IPO). The application can cover a single design or a set of related designs (for example, a tea set with the design applied to the cups and saucers).

For most UK design applications you need to complete and file official form DF2A. Notes on how to complete the form can be obtained from the UK-IPO Web site (see Appendix B) or you can ask the UK-IPO to send you the booklet 'How to apply to register a design'. The application includes the completed form DF2A and fee sheet (form FS2), one set of the illustrations (just one sheet if you only need a single illustration), and your fee.

The Designs Registry sends you or your representative a receipt, confirming all the details you provided and the date it received the application. Then it submits your application to an examiner, who issues a report with the results of the examination, normally within two months or so. If the examiner puts forward objections to the registration, he gives you two months or so to over-come these. If you're not successful in doing so, you still have certain rights of appeal.

If the application is successful the Registry grants the registration and pub-lishes the design – in the UK Registered Designs Journal and on the UK-IPO Web site. You can ask the Registry to defer publication, for example if you're not ready to market the product and wish to delay the design details becom-ing available to the public. Anyone who thinks that you're not entitled to the registration – for example, the owner of a similar existing design – can peti-tion the Designs Registry to remove the registration.

A successful registration runs initially for a period of 5 years from the date of application, and you renew it (on payment of renewal fees) for further peri-ods of 5 years up to a maximum of 25 years. As the owner of the registered design, you alone have the right to make, use, or sell the design for as long as the registration remains in force.

Providing illustrations

The illustrations that you send with the application must present an accurate and complete picture of the design. The illustrations can be photographs or drawings, or if your design is on a flat surface, you can give a sample. If the design is not two-dimensional or is merely decoration, include a series of views and label them (for example, Front view, Plan view, Perspective from above, front and one side). Use A4 paper, one side only, with the sheet number and sequence in the top right-hand corner (for example, Sheet 1/4, Sheet 2/4, and so on). You can get away, at least initially, with illustrations that don't meet all the Registry's formal rules but your 'informal' versions should still clearly show all the features you want to protect. The examiner asks for replacements if he's not satisfied with the first filed versions.

If any important design features aren't clear in the illustrations, you can include a brief explanation of these at the foot of Sheet 1.

To protect the design of just part of a product, clearly identify the part or parts by colouring, and/or solid lines for the part(s) and dotted lines for the rest, and/or circling the part(s) in red ink. Include wording that indicates which parts you want design protection for, and how you highlighted these in the illustrations.

Competitors assess the strength of your registration almost entirely on the illustrations. Try to ensure that they look good and display the design to good effect. If your illustrations look scruffy and carelessly drawn, the third party may see you as an easy touch and decide to infringe anyway. Unless you've good photographic or drawing skills it's probably worth paying for professional help. Most professional photographers are well equipped to help, and several excellent firms exist that are familiar with the drawings you need (see the list in Appendix B of some of those in the UK).

Applying for registration for the whole EU

You can gain registration for a single design across the whole of the EU. The registration, still known as the *Registered Community Design* (RCD), is a potentially very attractive option, requiring just a single set of application procedures and single set of fees. It offers a wide range of design protection, extending as far as signs and symbols that also qualify for trade mark registration. (See Chapter 8 for more on applying to register a trade mark).

The office responsible for granting EU design registrations is the awkwardly named Office for Harmonization in the Internal Market (OHIM), based in Alicante. This office is also responsible for granting EU trade mark registrations.

You can file an application for an EU-registered design at OHIM by mail or online at www.oami.eu.int. Alternatively, you can give the application form to the UK-IPO, which forwards it to OHIM for a handling fee (£15 on the 2007 schedules (fill out handling fee sheet FS3). You send subsequent communications and all fees, except the UK-IPO handling fee, directly to OHIM.

You can get your registered design application form online or as a hard copy from OHIM or the UK-IPO. You can complete the application form in any of the OHIM official languages (English, French, German, Italian, or Spanish) but OHIM requires that you also specify a second of these languages.

OHIM examine the application and advises you of any problems. They don't conduct official searches or novelty examinations of the application, but do check for basic formal requirements, such as including adequate drawings or photographs, and using the right forms and paying the right fees. If you fail to supply any of the required documents or information, you must submit these within two months.

OHIM classify designs into 32 classes – from Class 1 (foodstuffs), through such other classes as 8 (tools and hardware) and 21 (games, toys, tents, and sports goods), to 31 (machines and appliances for preparing food or drink), and 99 (miscellaneous). You don't have to specify classes when you make a design application; OHIM do that for you. Incidentally, classes 32 to 98 don't currently exist. Presumably OHIM have left them clear for adding new classes in future.

Registration and publication of the design typically occurs about three months after filing.

Like a UK national registration, the term of an EU design registration is an initial five years, and you can renew it for 5-year periods up to a maximum of 25 years, subject to the payment of renewal fees.

Applying for national registrations abroad

Within the EU you've the options of seeking design registration in all the EU member states by a single registration as discussed in the preceding section, or alternatively, applying by way of individual national applications in the EU members states (UK, Ireland, France, Germany, and so on). In general, if you have interests in more than two EU member states the EU route is less expensive than making national applications. Outside the EU the only route is via national applications.

Most countries have similar national systems to the UK for design registration, following a similar path to that described above in 'Applying for UK National Registration'. Applying by yourself to the local national IP office abroad may in some instances be possible but is not a route we recommend. You're almost certainly better served by enlisting the help of a local attorney to make the application and to take it through the steps to registration.

You can defer filing abroad by up to six months for design protection equivalent to that sought in a UK or EU application, while keeping the filing date of the UK or EU application as the 'priority date' of the equivalent applications. The IP offices in the countries concerned treat the UK/EU 'priority date' as effectively the date of filing in their country. (This is one of the provisions of the international Paris Convention – see Chapter 3.)

The way to protect your design in USA differs from that of other countries. The US Patent and Trademark Office awards design rights in the form of a design patent. The US design patent is valid for 14 years from the date of grant, with no renewal fees.

Considering Costs

A major factor in your decision on where to file applications for registered designs is likely to be the costs involved. Although the costs per design application are fairly modest, at least compared with the costs of gaining patent protection, a wide-ranging range of applications around the world can mount up to a substantial drain on the resources of your start-up business. The figures we quote in this section, based on 2007 fee schedules, are merely intended as a guide to assist you budgeting on IP costs. In all cases the level of attorney's fees vary depending on the amount of work involved in preparing the illustrations, and so on.

Registering designs in the UK

The application fee for a first design you include in a UK application is £60 if the UK-IPO gives immediate consent for publication, reduced to £40 if the UK-IPO delays consent. For any subsequent design you include in the application, you pay a further £40 if the UK-IPO gives immediate consent for publication, reduced to £20 if the UK-IPO delays consent.

Renewal fees for a UK registered design are

- £130 for the second five-year period
- £210 for the third five-year period
- £310 for the fourth five-year period
- £450 for the fifth five-year period

If you use an attorney, his fees for filing the design application may be about £300 to £600.

Registering designs in the EU

You must pay all the OHIM fees in euros. The official fee for an EU design application is in two parts: a registration fee and a publication fee.

The registration fee is

- €230 for one design
- €115 for each extra design up to ten
- €50 for each extra design above ten.

The publication fee is

- €120 for one design
- €60 for each extra design up to ten
- €20 for each extra design above ten.

You can ask OHIM to defer publication, for example if you wish to delay the design details becoming available to the public, by paying the following fees instead of the publication fee, but you must pay these with the application (and you still have to pay the publication fee when you do ask OHIM to publish the application!):

- ✔ €40 for one design
- ✔ €20 for each extra design up to ten
- ✔ €10 for each extra design above ten

Renewal fees for an EU-registered community design are

- ✔ €90 for the second five-year period
- ✔ €120 for the third five-year period
- ✔ €150 for the fourth five-year period
- ✔ €180 for the fifth five-year period

If you use an attorney you must add their fees to the above. Preparing and filing the EU design application may be about £500 to £1,000. For the various fee-paying stages, a rough guideline is that the attorney's charges are of the same order as the fee itself. But EU registration gives you such wide coverage and protection, that it's a considerable bargain.

Registering designs elsewhere

The best way to get an idea of the likely costs of registered design protection in other countries is to get your home country representative to provide an estimate based on his experience if the country in question. In broad terms, and including the charges of your home attorney and the attorney abroad, the costs are likely to be at least £500 to £1,000 for the filing. In countries with a full examination process, the costs of taking the application through to grant may probably be as much again.

Preserving Your Design Rights

Be vigilant about maintaining and enforcing your registered designs. It's your responsibility to make sure that you pay any necessary renewal fees on time. Maintaining your design registration is your responsibility and no one else's. Similarly it's you who's responsible for checking if any competitors offer products that infringe your design rights and to decide upon what actions you need to take in response. This section takes you through both aspects.

Maintaining your design registration

Renewal fees to maintain your design registration in force are generally payable every five years, to the IP office that granted the registration, on the anniversary of the filing date (the 'due date'). Late payment within a few months of the 'due date' may be possible on the addition of a penalty for later payment. Keep good diary records to make sure that you don't miss any renewal payments or face such late payment penalties. If you used attorneys to file and handle the application, their renewals department or agency should send you a reminder in good time before the due date, but keeping your own records too is a sensible precaution.

Enforcing your design rights

To enforce your design rights, you use similar methods to protecting your patent, so check out the section on patent enforcement in Chapter 6. You need to identify any instances of third parties offering products that appear to fall within your design registration, but having done so, check carefully that your protection is of sufficient scope to catch the offending product.

Dealing with what appears to be infringement of your rights makes good professional advice crucial. Even if you did not employ an attorney or solicitor in obtaining your registration, now is the time to enlist his aid.

Court actions tend to be long, very expensive, time consuming, and unpredictable. Pay particular attention to alternative means, such as negotiation and mediation, for resolving problems without resorting to court action. Your professional advisers can guide you on these alternatives, and Chapter 6 gives more information on agencies that specialise in them. You can find a list of the agencies in Appendix B.

Chapter 8

Applying to Register Your Trade Mark

This chapter talks about giving a trade mark to your product, service, or company, how to apply to register it, what the application process involves, and the likely costs.

What Is a Trade Mark?

A *trade mark* is a sign that identifies and distinguishes your goods or services from those of others. A more formal definition, from the UK Trade Marks Act (1994), is 'any sign which is capable of being represented graphically which is capable of distinguishing goods or services of one undertaking from those of another undertaking'.

A trade mark can therefore come in various guises. It may for example be formed of a word or words, letters, a phrase, numerals, a symbol, or a design. It may even be a colour, a sound, or in rare cases – a smell. (See Chapter 3 for a general introduction to trade marks.)

People often – and correctly – describe trade marks for services (as distinct from goods) as *service marks*. In this book, we use the words *trade mark* and *mark* to include service marks where appropriate.

A *registered* trade mark gives you the right to keep others from using it. A registered mark is an important part of your intellectual property; indeed, in the long run probably the most important and valuable part. As the trade mark's owner, you have the legal right – in the country in which you register it – to defend your company's mark for products or services against anyone using it without your permission on the same or similar goods or services.

You can use your trade mark without applying to register it – and importantly such use before applying doesn't harm a later registration application – but to secure it as part of your officially recorded intellectual property a registration is essential. An application to register also guards against the possibility of someone else registering the same or a similar mark before you do and thus having the prior rights to it. And if it ever comes to defending your rights in court you avoid the long, expensive, and potentially fruitless stages of showing that you'd acquired rights to use that mark.

The system that gives you rights to protect and defend your trade mark also works for other people and their marks. If you use someone's registered trade without consent, that person can sue you.

Looking at Examples of Trade Marks

Some of the most familiar trade marks can be found in the names of companies. They're often the names of a firm's founder, like Selfridges®; or of a favourite child (Mercedes® was the daughter of one of the car firm's founders). A trade mark may also be an invented word, like Tesco®. Initials are popular, for example BAA® and GKN®, but initials may be difficult to register because so many other organisations can lay claim to them. A similar problem arises with registering names of places, such as Durham® for carpets and rugs, which may need substantial evidence that the public associates the name with the applicant.

Under the umbrella of the company name it's also commonplace – and indeed very useful – to have a registered name for the services or products on offer. Motor companies make great use of this: Ford® Galaxy®, Nissan® Micra®, Toyota® Corolla®. Registering product names like this combines the reputation of the firm as a whole with the properties and reputation of the individual model. Many companies combine their name with a logo: NatWest Bank's three arrows with a triangular outline, Lloyds TSB's black horse, or McDonald's golden arches.

To complete what can be an impressive array of IP rights, a company may have trade mark rights in a slogan (Coca Cola's 'It's the real thing'), a colour (Cadbury's distinctive purple on its chocolate bars), music (remember Hamlet Cigars' use of 'Air on a G String'?) or a shape (Coca Cola again, with their distinctive bottle shape). A number of companies have even succeeded in registering a smell as a trade mark (Unicorn Product's beer odour for dart flights) but registration of a smell is very difficult given the need to present it 'graphically' (see the definition of a trade mark at the beginning of the chapter).

From a marketing viewpoint, combinations of different kinds of mark are understandably very popular, with the name, logo, colour, jingle, and shape all adding to the public's identification of the brand.

Choosing Your Trade Mark

Your aim in choosing a trade mark is to help customers to make repeat orders of your product or service. Faced with a choice between a known and trusted brand (a 'Bosch®' dishwasher) and an unspecified or unknown brand (a 'Blotto™' dishwasher), the likelihood is that most customers opt for the known and trusted version. Choose a mark that's readily recognised and easy to remember, but do avoid one that too closely resembles someone else's mark for the same or similar goods or services.

For your new business start by protecting a name for the company and the product (in the first years the same name for both may suffice), possibly in combination with a logo. Slogans, smells, and signature tunes can probably wait for now. Try to come up with a name that stands out from the crowd. But also seek a word or words that sound distinctive, easy to remember, and easy to say.

Don't mix up the purpose of your core trade mark – to identify and distinguish your goods or services from those of other suppliers – with other important elements of your marketing. Many inventors try to make the trade mark a triple whammy: not only the identifier but also a description of the goods or services, and their desirable qualities. The message is much stronger if you separate out these three elements, as follows:

- ✔ Trade mark: L'Oréal
- ✔ Description: Eye contact cream
- ✔ Slogan: 'Because you're worth it'

And take care not to make the words too descriptive of what the product or service does. If you do so, you make the trade mark more difficult to register and may prevent valid registration altogether. Trade mark registries and the courts aren't keen to grant or uphold registration of descriptions that everybody else should be free to use.

Similarly, avoid names that have no distinguishing features other than being in an unusual typeface (though note that a new typeface as such may merit design protection – see Chapter 7), or are simply misspelt (fone for phone), presented as an Internet domain name (xxx.com or xxx.co.uk), offensive, against the law, or deceptive.

The easiest marks to register are invented words like Kodak®: clear, easy to remember, and unlike anything else.

Building a brand with a trade mark

The essence of branding is to build a reputation in the minds of potential consumers. The branding power of your trade mark lies in its ability to lure potential consumers towards your product or service. If you put your brand name on a package it may be interesting and informative, but if it also sticks in the consumer's mind the result should be a larger market share for you.

A new brand name must generate positive publicity or it won't survive in the market. If you want to be successful in influencing the minds of potential consumers, build a distinctive brand that appeals to your customers more than your competitors' brands.

Asked to buy tissues, for example, most people automatically think of Kleenex – by far the leading brand. Even if they look across the room and see a box of Scott tissues, they may still ask, 'May I have a Kleenex?' Likewise, people tend to use to Xerox as a generic term for photocopiers and Hoover to represent all vacuum cleaners.

As your brand becomes established and better known you must avoid it becoming so commonly used that it becomes the name of the product. In trade mark terms, your mark then becomes *generic*, and you lose your exclusive rights to it because anyone can use it for such a product. Examples include nylon and escalator: both used to be registered trade marks. You can maintain your exclusive trade mark rights through proper use only (as an adjective: *'Bisto'* gravy, not just *'Bisto'* as a noun) and rigid enforcement when someone uses your mark incorrectly.

Notice that if you ask for a Coke® in a bar that stocks Pepsi® the bartender is likely to say, 'We sell Pepsi. Is that alright?' The Coca-Cola company put huge effort into making sure that its immensely valuable brand does not become generic, even to the extent of monitoring the way bar staff operate.

Checking what's already out there

Having decided on a possible mark, and ideally a few reserves, you need to check whether somebody else has beaten you to it and registered those marks for the goods or services you offer, or plan to offer. Little point exists in using a mark if an existing owner is likely to sue you for infringing it. Likewise there's little point in applying to register a mark if someone else has already registered it. The trade mark examiner tells the owners of possibly conflicting registered marks about your application, to check whether they object to yours joining theirs on the register.

You can search for prior registrations yourself or enlist professional help to do so. Using a trade mark attorney to handle your trade mark application may or may not be essential, depending on the country or region you want to cover. But, as with patents and registered designs (see Chapters 6 and 7), we strongly recommend enlisting such professional help. This is especially so with international applications. Have a look at Chapter 4 for more details on finding a representative.

Using the trade mark symbols

You may use the ™ (trade mark) symbol to alert the public to your claim to ownership of a mark, regardless of whether you file an application to register it. Doing so puts others on notice that you regard the mark as your property and indicates that you're thinking about, or are in the process of, registering the mark. But at the same time it indicates to your competitors that you've yet to gain registration of the mark: if you'd successfully registered the mark you'd surely be using the registration symbol ®.

You may only use the registration symbol ® *after* you successfully register the mark, *not* while the application is pending. Also, you may use the ® symbol with the mark only on or in connection with the goods and/or services for which it's registered.

The registration denoted by the ® need not necessarily be in your home country but be careful not to mislead the public by stating in words that the mark is registered in a given country when it isn't.

Filing Your Trade Mark Application

This section deals first with applying for a UK trade mark registration. Most other national applications have similar requirements and we also cover applying for an international registration under the *Madrid Protocol* and applying for an EU Community Trade Mark. See Chapter 3 for the basis for these international arrangements.

Before you file the UK or other national application, pause to consider the extent of your market in terms of countries that may want your product or service. If you have interests in more than two or three EU countries it may be cheaper and easier to cover them all (including the UK) in a single EU registration. If your interests extend even farther afield, then based on a national application or registration you can make a Madrid international application that covers as many of its member countries as you want.

Applying for a UK national registration

For preparing a UK application we strongly recommend that you read the UK-IPO booklets 'Trade marks: essential reading' and 'Applying to register a trade mark'. Both are available from the UK-IPO Web site (www.ipo.gov.uk).

In the UK, you apply to the UK-IPO to register a trade mark, using form TM3. You must fill in Section 2 (illustration of the mark or marks), Section 9 (the list of goods or services you want the registration to cover), and Section 12 (your full name and address), and you must sign the declaration at Section 14. Ensure that you send the official fee with the form (see the later section 'Looking at Costs' for fee details) and the official fee sheet, FS2.

Your details, including your name and address, appear in due course on the open records of the UK-IPO and in the *UK Trade Marks Journal*. If you don't want your home address published, provide a different address or a PO box number.

The UK-IPO records the mark exactly as you present it in the Section 2 box or, if it's too big to fit the box – and assuming you're not later aiming to file an international Madrid application – on a separate sheet up to A4 size. Don't include underlining, quote marks, and so on unless they form part of the mark. And if the mark is a logo or picture, don't include words that aren't part of it.

Listing goods and services

Use clear, concise, easily understood terms for the list of goods and/or services you want to cover. Also consider the official class or classes in which those goods or services fall. The UK-IPO have 45 classes for trade marks ranging from Class 1 (chemicals used in industry) to Class 34 (tobacco and so on) for goods, and Class 35 (advertising and business management) and Class 45 (legal services) for services. You can find a full listing of the classes and their contents on the UK-IPO or World Intellectual Property Organisation (WIPO) Web sites (www.ipo.gov.uk or www.wipo.int).

Delivering the application

You can take the application papers by hand or post them to the UK-IPO (the address details appear in Appendix B) or you can apply online using the electronic version of form TM3. The Trade Marks Registry sends you an official receipt of filing and passes the application to an examiner to check for any objections to registration.

Proceeding through the Trade Marks Registry

A major potential source of objections is an existing registration for the same or similar marks for the same or similar goods or services. In the case of such a prior registration, the Registry sends details of the application to the owners of that registration, to give them the opportunity to object to your application.

If the examiner or a prior owner raises no objections, or if you can overcome any objections that they do raise, the Trade Marks Registry advertise the application in the *UK Trade Marks Journal*. The advertised application provides a period of three months in which anyone else can object to the registration. Again, if no one objects or you can overcome any objections, the application proceeds to the next stage: registration of the mark and the issue of your trade mark registration certificate. Well done!

Renewing the registration

The UK registration runs for ten years from the date of application. You can renew the registration for further periods of ten years, again and again, subject to payment of the official renewal fees. The possibility of keeping the registration in force *for ever* makes it extremely valuable. So make sure that you or your attorney have in place a good diary system to avoid missing the renewal payments.

Applying for an international registration

After you have an application or registration for your mark in your home country you can apply for a single international registration that covers any other member countries of the Madrid Protocol. This is an international agreement under which you, as the owner of a trade mark in one of its member countries, (as an application or granted registration) can protect it by an international application that covers other member countries. See Chapter 3 for more on the Madrid Protocol.

The Madrid Protocol comprises some 80 member countries in total, including the UK and all the other EU member states, Australia, Japan, and the USA. (See the WIPO Web site for a full list www.wipo.int.) You can designate as many of the countries as you want, and even after registration you can add more (using WIPO form MM4).

You apply for the international application on the application form MM2(E) – available from the WIPO Web site. Many of the details you enter on the form must be identical to those on your home country trade mark, including your name and address, and the illustration of the mark (which for a Madrid registration must not exceed 50×50 millimetres). The list of goods and services must be no wider than for your home mark.

The application must include the completed and signed form MM2(E), plus form MM17 if you designate the EU, and form MM18 if you designate the USA. Application fee payments depend on the countries you designate and you pay them later to WIPO. See the 'Looking at Costs' section, later in the chapter, for more on fee payments.

Delivering the application

You file the international application at your home IP office. For the UK, take or send the application to the UK-IPO (details in Appendix B), with the envelope marked 'For the attention of the Madrid Application Unit', or you can fax it to the UK-IPO (01633 817777) but the original copy of form MM2(E) must follow by mail. You must accompany the application with a UK handling fee unless you'll be filing a subsequent designation of countries (on form MM4). The official filing date is the date the UK-IPO receives the application.

Proceeding through the WIPO stages

The home country IP office (in the UK, the UK-IPO) forwards the application to WIPO in Geneva for processing, sending to WIPO the application papers, and a certificate to show that the international application is for the same mark as in the home country. WIPO examines the application for any omissions, collects the fees, and registers the mark. WIPO then reports the registration to each of the designated countries, which in turn examine the mark under their own trade mark application procedures. If a country raises objections – for example, a conflicting registration exists there – their IP office must advise WIPO, which forwards details to you. You can then reply directly to the country's IP office to attempt to overcome the objections.

The international registration lasts for ten years from its date of registration and you can renew it indefinitely for further ten-year periods, subject to the payment of renewal fees.

Applying for an EU trade mark

You can apply for an EU *Community Trade Mark* (CTM) that covers all the EU countries in a single registration. The Alicante-based office for the Harmonization in the Internal Market (OHIM) handles CTM applications (we give you contact details in Appendix B).

You can download the application form from the OHIM Web site – see Appendix B. The information you enter on the form is similar to that for national applications: name and address of applicant, details of the mark, list of goods or services to cover, representative's details (if used), and any claim to give the EU application the priority date (the same effective filing date) of an equivalent application, say in the UK, filed within the previous six months. You must indicate on the form the language in which you're filing and also select a second language from the five OHIM languages (English, French, German, Italian, Spanish) to be available for any proceedings in which a third party may object to the registration.

Delivering the application

You can file the application directly at OHIM, by mail, online, or indirectly via your EU home country IP office (in the UK, the UK-IPO). Indirect filing via the UK-IPO incurs an additional handling fee.

EU residents can employ a representative if they choose – but applicants from outside the EU must use one. The representative must be on the OHIM-approved list. We strongly advise you to use a representative, even if in principle you can represent yourself.

Proceeding through the OHIM stages

To achieve registration, the mark must be distinctive and free from conflict with prior rights of others. OHIM examines the application to check that it's formally correct and undertakes a search for prior registrations. It also arranges a check for prior registrations in other EU member countries. With the search results assembled, OHIM reports to you or your representative any earlier identical or similar CTMs and CTM applications for identical and similar classes of goods and services, together with national search reports from other member countries.

About one month after the search report OHIM publishes the application and at the same time OHIM also advises the owners of the CTMs in its report, to give them the opportunity to object to the registration.

If no objections arise, or if you overcome them, OHIM invites you to pay the registration fee. OHIM formally records the registration on its database after you pay the fee. If OHIM don't receive payment within an initial two-month period from the invitation, or within a further two months on payment of a surcharge, it deems the application withdrawn.

A CTM registration lasts for ten years from its date of registration and you can renew it indefinitely for further ten-year periods, subject to the payment of renewal fees.

Looking at Costs

This section covers costs for registering trade marks in the UK, internationally under the terms of the Madrid Protocol, and within the EU. The figures we quote relate to 2007 fee schedules.

UK-registered trade marks

The application fee to register a trade mark in the UK in one class of goods or services is £200, plus £50 for each extra class.

Likewise with renewal fees £200 for the first or only class of the registration, and £50 for each extra class (submitted to the UK-IPO with form 11). On receipt of the payment, the UK-IPO sends you a renewal certificate, confirming your mark's renewal for the next ten-year period.

If you use an attorney, her charges for filing the UK application may be about £300 to £600, but vary according to the amount of work involved in preparing and filing the application. Application fees for national applications in other countries are of a similar order, and remember to factor in the local attorney's charges.

Madrid Protocol international registered trade marks

You pay the official fee for an international trade mark application to WIPO in Swiss francs. The fee depends on the mark, the number of designated countries, and the individual designation fees of those countries. In total, you pay:

- A basic fee of SFr 653 (plus an additional SFr 250 if the mark is in colour)
- A standard designation fee of SFr 73 for each designated country or the amount the country concerned designates
- A supplementary fee of SFr 73 for each class of goods and services after the third class (the basic fee cover up to three classes)

Madrid applications filed via the UK-IPO incur an official UK handling fee of £40.

The renewal fees are similarly:

- A basic fee of SFr 653 (subject to a 50 per cent surcharge for late payment)
- A supplementary fee of SFr 73 for each designated country in which you must pay individual fees
- A complementary fee of SFr 73 for each designated country in which you don't pay individual fees

If you use an attorney you must add her fees to the above costs. Preparing and filing the Madrid application may be about £500 to £1,000. You also pay a handling charge for attending to the payment of renewal fees.

EU Community registered trade marks

The official fees for an EU application for a single mark are:

- ✔ A basic fee of €900, reduced to €750 for online filing
- ✔ An extra fee of €150 for each class of goods and services over three (the basic fee covers up to three classes)

An application filed via the UK-IPO incurs an official UK handling fee of £40.

The renewal fee for an individual registration is

- ✔ €1,500, reduced to €1,350 for online filing
- ✔ An extra fee of €400 for each class of goods and services over three

Attorneys' fees for preparing and filing an EU CTM application are likely to be of the same order as those for a Madrid international application. Similarly, you pay a handling charge when renewing.

Enforcing Your Trade Mark Rights

Our advice on all matters concerning enforcements of IP rights, including trade marks, is the same: Don't rush at once to the courts. Instead, check out carefully what your rights are and what the suspected infringer is getting up to. Contact your attorney immediately and get her advice on the possible and recommended courses of action. Wherever possible, reach a settlement by negotiation with the third party or their representatives. And if doing so fails, make sure that in any court action you have good representation.

Finally, remember that for court actions over trade marks you can choose between civil and criminal actions, and that across Europe the latter offer good prospects of a successful outcome for the IP holder and substantial grief for the infringer – financial penalties and custodial sentences. Chapter 6 on patent enforcement and Chapter 9 on copyright enforcement have more to say on IP actions and negotiating settlements.

Chapter 9

Using Your Copyright to Stop the Pirates

Copyright is an often misunderstood and sometimes overlooked branch of intellectual property (IP), despite – or perhaps because of – being automatic and free.

While putting in all the hard graft designing a new product, and then getting ready to market and sell it, do keep in mind the useful copyright – legal rights against copying – that arise as you create not just the product design, but also any related artwork, instructions, and promotional material.

This chapter explains what copyright is, its advantages, and how long it protects you.

What Is Copyright?

Copyright is a right for the creator of an original work to take legal action against unauthorised copying of the work.

Types of work that may possess copyright include those from such a person as an author, composer, designer, motion picture maker, musician, painter, playwright, or photographer.

The UK Copyright, Designs and Patents Act (1988), defines copyright more formally:

> *Copyright is a property right which subsists in original literary, dramatic, musical or artistic works, sound recordings, films, broadcasts or cable programmes, and the typographical arrangement of published editions (of the whole or any part of one or more literary, dramatic or musical works).*

Recognising Your Copyright

Copyright exists in your technical notes on your product, in its technical drawings (which include the drawings in your patent specification) and in your related computer software. Your notes, drawings, and software very likely make up the most valuable parts of your copyright, but copyright also exists in the packaging or displays you're going to use to bring your product to market: you may need instruction leaflets, posters, written messages on the packaging, DVDs or CDs explaining your product, or promotional material you give to suppliers, wholesalers, and distributors.

All of this copyright can give you valuable advantages over your competitors. Unless they have your approval they can't offer a product that copies your design: they must create their own. Without your approval they can't copy your software or your packaging, or take the wording of your instruction leaflet. Preparing alternative, non-copying versions of all these elements takes up your competitors' time while you forge ahead with production and marketing.

The copyright laws treat the protected categories of copyright broadly. So, artistic works include maps and architectural plans. Dramatic works include any accompanying music. Films include motion pictures and other audiovisual works. Literary works include computer programs.

Copyright not only protects you against copying of the work, but also against someone adapting it, distributing it, communicating it to the public by broadcasting, renting or loaning it to the public, or performing it in public.

Converting someone else's computer program into a different computer language counts as adaptation, and thus copying. Sending someone else's copyright material over the Internet is copying, and similarly, downloading or distribution of copyright material that others have put on the Internet is copying. Even photocopying a newspaper article is copying.

Copyright also protects a database if the selection or arrangement of its contents is original. The database may also acquire a *database right* if its preparation involves substantial investment. Database right lasts for 15 years from when you make or publish the database and protects against unauthorised extraction and use of its contents.

Although no official copyright register exists in most countries, because copyright there is automatic, a number of companies offer *unofficial* copyright registers. Some of them write to you in a manner that suggests registration with them is essential – in return for a fee for their trouble. Think very carefully about whether entry in an unofficial copyright register is a useful service for you.

The US is unusual in having an official system for registering copyright, but doing so is optional: The work in question still possesses automatic copyright even if you don't register it. Unless you have unusual and particular needs in the US market, you need not be unduly concerned about registration.

Understanding what copyright does not protect

Material generally not eligible for copyright protection includes the following:

- ✔ Works that aren't recorded in writing or any other way (for example, choreographic works or improvised performances that no one records).
- ✔ Titles, names, short phrases, and slogans.
- ✔ Familiar symbols or designs.
- ✔ Mere lists of ingredients or contents.
- ✔ Ideas, procedures, methods, concepts, principles, or discoveries.
- ✔ Works consisting entirely of information that's common property (for example standard calendars, height and weight charts, tape measures and rulers, and lists or tables taken from public documents).

Knowing who can claim copyright

Unless otherwise arranged, the copyright in a creative work is the property of the author or authors who create it.

But, as with any other kind of property, the author(s) of a creative work can sell, bequeath or otherwise assign (transfer) their copyright to another person or organisation. The receiving person or organisation becomes the rightful owner of the copyright. For example, an employer is the copyright owner of work created by an employee in the ordinary course of the employment. The author(s) do however retain the right to be identified as the authors.

If you subcontract such aspects as software, packaging designs, and promotional DVDs to specialists in these areas, make sure that you understand who owns what. Get suitable agreements drawn up and signed, if at all possible *before* the creative activity starts, with the aim of indisputably making you the owner of all the copyright involved. Otherwise when your product starts making money, the owners of copyright that hasn't been assigned to you can sue you for infringement of their copyright. See Chapter 11 and the 'Transferring Copyright' section below for more on hiring subcontractors and for ensuring that you don't lose control of the copyright.

For the avoidance of future disputes, joint authors of a creative work should similarly agree in advance their respective shares of any remuneration or other rewards that may arise from the work or its assignment to a third party.

Mere ownership of a book, manuscript, painting, or any other copy or sound recording doesn't give the owner the copyright. For example, if you buy a CD with music on it, you own the CD, but you don't own the rights to copy what's on the disk or to sell it to others. The artist or record company owns the rights to the music. You would need permission to make copies.

Using other people's copyright

As with many aspects of IP you need not only to be mindful of your rights but also to take care that in going about your business you don't breach the rights of other people. To use other people's copyright work you need their permission, in some cases obtained or negotiated directly with them and in other cases with an organisation that acts for groups of copyright owners. Among these groups is the Newspaper Licensing Agency (NLA – you can find their contact details in Appendix B), which licenses business users to photocopy, scan, and email articles from national, local, regional, and foreign newspapers. Another is the Performing Right Society (PRS – again, their contact details are in Appendix B), which collects licence fees from music users to be paid to the music's writers and publishers.

Certain actions with regard to copyright material *are* acceptable. In the UK, limited copying of a work for non-commercial research or private study is generally permitted. So is quoting from a work for the purpose of criticism or review – provided that you accompany the quote with a sufficient acknowledgement. Simply using part of a work (other than a photograph) for the purpose of reporting current events is permissible, again provided that you include a sufficient acknowledgement. And time-shifting of TV programmes by recording them at home for later viewing at home is allowed. But still be careful! If you copy large amounts of material or make lots of copies you may still need permission from the copyright owner.

Safeguarding Copyright Protection

Let's first repeat the welcome message that copyright is *automatic* and *free*, not only in the UK but also in most of the rest of the world. You don't need any publication, registration, or other official action to secure copyright. However in the case of disputes or actions to enforce your copyright you have to show just what you regard as your copyright material and have to supply sufficient evidence to prove that you're the true owner.

Recording your copyright

The primary aim of recording your work is to build up the evidence to show your ownership of the work at a date you can prove in any dispute. If you can clearly show your prior ownership, there's a good chance of the competitor who's selling a copy backing off with no further argument. And even if the dispute does get to court, you've already assembled a key part of your evidence.

So as you beaver away with your creation, never forget the rights you're acquiring as you go. That way you're more likely to make a decent record at each stage of its development. See the section in Chapter 1 on keeping good records. A dated logbook helps you prove the dates at which the work in question was made and completed.

For all 'visually perceptible' work, in other words, work capable of being seen by a reader or person viewing it, make sure that you mark it as copyright. The standard way of indicating copyright material is to include the international copyright symbol, © (or the word 'copyright' or the abbreviation 'copr'), followed by your name and the year of first publication, for example © John Smith 2008. This way is especially suitable and desirable for industrial copyright in your technical drawings, brochures, leaflets, and so on. But it's also important to include the date of production of the work. Technical descriptions and technical drawings in particular should carry the precise date of creation in addition to the standard © mark.

Copyright marking warns your competitors that you're well aware that copyright protects your material. You send a clear message akin to 'Trespassers will be prosecuted'.

Part of the popular folklore on copyright is that you should record your ownership at a provable date by sending yourself a copy of the work in a sealed envelope and by recorded delivery – to get an official date stamp – and then not open the received envelope. This undoubtedly makes the sender feel better but may not provide conclusive proof. How can the court be sure that the revealed contents of the envelope are those that were in it at the stamped date? Easy-seal envelopes can be opened and easily resealed; just as gummed ones can be steamed open and stuck down again.

A better variation on the folklore is to send the work (in confidence) to your solicitor or bank manager. The intention is to get them to confirm at a necessary stage, say by a sworn statement, that the work was known to them the date they received it.

Using international copyright

Most countries automatically protect your copyright to a similar, if not always identical extent, to UK laws, under certain international IP arrangements. The four arrangements of most relevance to copyright are as follows below (see Appendix B for contact details). In all four the contracting states include the UK and all other EU member states, the US, China, and Japan and so on. You can find further details of these and other IP treaties and conventions in Chapter 3.

- Berne Convention for the Protection of Literary and Artistic Works.
- Universal Copyright Convention.
- Rome Convention for the Protection of Performers, Producers of Phonograms and Broadcasting Organisations.
- Trade-Related Aspects of Intellectual Property Rights (TRIPs).

Some non-governmental organisations can also be highly effective in promoting copyright and encouraging governments to put in place and enforce the necessary IP legislation. The Motion Picture Association of America achieved notable successes in persuading the governments of certain countries in the Far East to introduce and enforce restraints on IP piracy.

Calculating copyright duration

Although regularly renewed trade mark registration can last forever, copyright protection can in practical terms be very long lasting too. The periods vary according to the nature of the work and to some extent from country to country. Table 9-1 shows the current periods of protection relevant to the UK.

Table 9-1	Copyright Protection Periods in the UK
Type of Work	*Duration*
Literary, dramatic, musical, artistic, photographic, recorded speeches, computer programs	The author's lifetime plus 70 years after their death
Films	The lifetime of the principal director, screenplay or dialogue authors, or of the film's score composer, plus 70 years after that person's death
Broadcasts, sound recordings	50 years from publication
Typographical arrangement of published editions	25 years from publication
Database right	15 years from making or publishing the database
Three-dimensional designs (UK)	15 years from the end of the first year in which a design document recorded the design, *but reduced to* 10 years if marketing occurs within 5 years from that first record
Three-dimensional designs (EU)	3 years from the first availability of the design to the public

Enforcing Your Copyright

As with other forms of IP, don't start throwing out claims of infringement (writs), the moment you think that someone is using your copyright work without your permission. Check out what that person's doing with your material and how closely their work resembles yours.

More than in other kinds of IP, any action involving copyright almost invariably requires professional input. There's no official record of your copyright (except possibly in the USA – see the 'Recognising Your Copyright' section earlier in this chapter). So help is desirable from the outset to establish good evidence of the existence and extent of your rights. In due course this evidence may need to be presented to third parties and the courts. Patent attorneys and specialist IP solicitors can help here. See Chapter 3 for more on who they are and how to choose them.

Several organisations help IP owners enforce their rights. In the UK, such organisations include the Federation Against Copyright Theft (FACT), Anti Copying in Design (ACID), the Federation Against Software Theft (FAST), and the Alliance Against IP Theft. See Appendix B for their contact details.

Your first step towards resolving a copyright issue with a third party is an attempt to settle the matter by negotiation. Your professional helper can put together letters and approaches that avoid – or at least reduce the prospect of – angry exchanges that make matters worse. See Chapter 6 for more on using mediation or other forms of alternative dispute resolution that aim to reach a settlement without resorting to the courts.

If the infringer is a reputable company – large or small – you may find it easier to achieve a negotiated settlement. If not, the most appropriate action may be a civil court. The most difficult disputes to resolve can be with unscrupulous street market traders dealing openly with your pirated goods. In these cases you may prefer to take action in a criminal court. Across the EU the criminal provisions against pirate traders are particularly strong, with the prospect of a custodial sentence for a party found guilty. Your local trading standards officers may also be able to help. Many councils have such officers who specifically tackle pirate traders.

Although offering strong protection if it comes to court actions, copyright does place upon you a strong burden of proof, especially in the civil courts. You have to convince the judge that you own the copyright, that the offending competitor's offering is sufficiently similar to yours, and most importantly that the competitor truly copied you. If the competitor can show that he did not know of your work and created it independently then there was no copying, so there was no copyright infringement.

Transferring Copyright

As an inventor, you're likely to need to arrange a transfer (assignment) of copyright at some point. As discussed above, you may want to ensure that the copyright in work you pay a subcontractor to undertake is confirmed as your property. Until you get the person with the copyright to sign a suitable agreement they still own the copyright. To be completely sure that it's properly transferred to you get the subcontractor to sign the right over to you before they start work (see Chapter 11 and the example in Appendix A of an 'Assignment and Work for Hire Agreement').

Similarly if you decide to sell to a third party (for we trust a good return to yourself) your invention and its associated IP rights, including copyright, then you need to sign a suitable assignment to the third party.

The transfer of copyright from its owner to you or another agreed party isn't valid unless that transfer is in writing and signed by the owner of the rights conveyed, or the owner's duly authorised agent. If the transfer is signed on behalf of an organisation, make sure that the person signing is a properly authorised representative.

Copyright is a personal property right, subject to the various laws and regulations that govern the ownership, inheritance, or transfer of personal property, as well as terms of contracts or conduct of business. You can if you wish leave someone copyright in your will. Which is a suitable point at which to end this chapter.

Part III
Developing Your Idea

'So I told the weedy little inventor of the Little
Brother Domestic Robot – "Do your worst –
we'll not pay you a penny for your invention!"'

In this part . . .

The journey from brilliant idea to world-beating product is a complex one. Visualisation, prototyping, testing, and market research are all key parts of that process.

The chapters in this part of the book take you through these development steps, both for new products and for improved processes.

Chapter 10

Showing That Your Invention Works

*Y*ou probably believe that your new invention definitely works, but you can't actually tell until you test it out. A world of difference exists between having a great inventive idea, however groundbreaking, and putting it into practice.

Reaching a developed, marketable version of your invention depends to a great extent on what kind of invention you create. Developing a new mechanical or structural device is very different from producing a new washing-up liquid. And the invention may not be for a product as such, but for an improved way of making it, or for new software to operate it.

The field for many first time inventors is however that of new mechanical or structural items: a workbench, exercise machine, can opener, clothes airer or kitchen stool. A crucial part of the development of these inventions is the production of a *prototype*: a working model (sometimes called a *proof of concept prototype*) that provides functional proof that the invention works. Therefore, we devote much of this chapter to designing a working prototype model of a new mechanical or structural product. We explain what a prototype is, tell you all about the advantages of prototyping, and walk you through the process of turning your concept into a practical form.

The needs for developing and proving an improved process or composition generally follow the same lines as those for making a mechanical or structural product, and we comment on where they differ markedly from those of making a prototype. If your invention is a new process you need to experiment with it and confirm that it gives the improvements you expect. And if your invention is a new composition – like a new toothpaste – you need to produce some samples to confirm that it actually does what you claim.

Understanding the Importance of Showing Off

People love to see a new idea demonstrated. They want to see that the idea works or, in the case of a totally new concept, at least possesses the potential capacity to work. Having a product prototype helps convince people of the merits of your idea. For a process, you can produce a DVD or video that shows it in operation and select good samples of the product of the process to show off. For compositions, you can again have selected samples, including some that interested parties can try out.

Consider the following:

- ✔ You can demonstrate a prototype, DVD, or samples to private investors and bankers when seeking funding for further development.

- ✔ Because companies tend not to license mere ideas, plan to show your demo material to potential licensees to encourage them to license your invention and pay you for it. (See Chapters 21 and 22 for more about licensing and negotiating a good deal.)

- ✔ You can use your demo material to impress suppliers and wholesalers so that you can establish lines of credit.

Having a prototype, DVD, or samples in hand helps those who see them grasp and assess the idea that you're trying to describe.

Turning your ideas into practice presents you with a great learning curve. The devil is in the detail of turning your concept into reality. Solving how to make the thing perform well in practice often leads to further good inventions within the scope of the original concept. Indeed, more detailed inventions often provide the most valuable parts of your patent rights (see Chapters 5 and 6 for more on how to secure and defend your patents).

Developing Your Idea

You can take your idea forward in one of three ways:

> ✔ If you have the talent and the facilities, you can develop it yourself.
>
> ✔ You can hire a professional designer or researcher to develop it.
>
> ✔ You can work with an existing company that's active in your invention's marketplace, getting the company to study your concept and then develop it.

Most people aren't able to go it alone, so don't feel bad if you have to get some help. This chapter provides an overview of the development process and Chapter 11 covers hiring others to help you.

A wild idea can often lead to a mental vision of a new way of doing something or possibly to a new machine that makes doing a task more efficient. Before you can convincingly convey your vision to someone else, you need to make or obtain reasonable descriptions or sketches, firstly to satisfy yourself that your idea works, and secondly to allow others to understand what you aim to do.

Obtaining Your Prototype

This section discusses in detail the options for creating a prototype of a mechanical or structural invention.

Going through the process yourself

If you have an extensive knowledge of the prototyping process, and you have the skills and facilities needed to complete each step of the process, then by all means do it yourself. But whether you decide to create your own prototype all by yourself, or you hire someone to help you with it, you go through the following phases.

Paper, pencils, and rubbers cost much less than the materials and fabrication work you require to make a proof of concept prototype. The better your pre-prototype drawings, the greater the chance of making an exceptional proof of concept prototype.

> # Rapid prototyping
>
> 3D computer-aided design has now advanced to the point where you can programme a machine that creates a near instantaneous 3D image or even a model from a CAD drawing. Then you can make an actual component part or small non-functioning prototype or model. Although computer technology may be expensive, it requires only hours to provide functional prototypes that otherwise take days, weeks, or months to accomplish manually. For example, if you want to make a kitchen blender, it may take a month of manual labour to fabricate certain component parts from chunks of metal or plastic. Rapid prototyping can cut the labour time to little more than a day.

Technical drawings

After you complete your basic pre-prototype sketches, the next phase is to move on to a drawing board or, better still, to *CAD* (computer-aided design) software to refine and improve the design. As you work, keep in mind that your invention needs a logical, functional, and economic manufacturing process. Your technical drawings also reveal the various functional states of the expected operation, to demonstrate how a consumer uses your product.

Your technical drawings need not involve any aesthetics, nor need they cover the final colour, size, shape, and packaging. You can worry about those details later. And your wholesalers, market outlets, and prospective licensees will welcome having a say in the final design.

Study model

After the design drawings demonstrate that your concepts may satisfy an existing need, you need to make, or employ someone to make, a prototype model that demonstrates your product's usefulness and market appeal.

Design rendering and graphic drawings

After the prototype reveals that the concept can function reliably and well and has market potential, you can now think in more detail about the aesthetics of your product's design.

Eye appeal sells products, so don't take the aesthetic design step lightly.

After the final aesthetics determine your product's form, complete detailed graphic drawings and show these to potential users before you commit to funding the expensive final preproduction prototype. Drawings can be changed much more easily than tooling. Don't reject lightly any modifications that potential users suggest – they may give you a better understanding of the market's likes and dislikes – but equally you must satisfy yourself that the modifications can be justified.

Preproduction model(s)

After you test your proof of concept prototypes and gain input about the functionality, aesthetics, and market likes and dislikes, you use this information to design your final production model prototype. The resulting prototype is the one that you show to potential buyers, investors, or licensees.

Hiring a model maker to do it for you

Prototyping typically involves creating several proof of concept models that lead up to a *market-ready version*, which is a preproduction model that incorporates all the mechanical and aesthetics of the proof of concept prototypes. In working with a professional model maker you travel through each of the stages at an advanced rate, which saves you aggravation and time.

A professional model maker can create your prototype for you. You may hire a local engineering company or an engineering pattern maker. You may also, or instead, want to contact the engineering department of a local university to obtain technical assistance. Universities and engineering schools have workshops that can undertake the work, or point you to tried and trusted professional prototype makers who can. Local inventor organisations in your area also provide great resources for prototype connections. Appendix B includes a list of support organisations that may be able to help or point you to others who can.

Prototype makers tend to specialise in different industries. For example, if you have a medical product, seek out a product designer who specialises in medical products. Your specialist designer is probably already aware of the government rules and regulations in the medical industry. The designer knows that you can't use specific plastics, resins, and chemicals at all in the development of a new medical product, but is aware of other materials that provide advantages you may never have thought of.

When looking for a prototype designer, select one with a proven track record. Check out the list of successful products the designer has worked on. Ask for references and thoroughly check them out. An experienced designer not only has the ability to make design suggestions, but also has contacts with manufacturers that may assist you in further development, and possibly production and licensing.

Don't think that you have to make only one prototype. As the product develops, it changes. Many inventions require several changes during their development, often based on recommendations from buyers, wholesalers, and consumers. Inventors commonly have several different prototypes as the product develops.

James Dyson made over 5,000 prototypes before settling on the design for the first bagless vacuum cleaner.

Assessing Your Prototyping Needs

Prototyping can be very simple or quite complex depending on the nature and type of the product design. Each prototyper has an area of expertise, which may be primarily electronic, mechanical, chemical, and so on. Prototyping companies often specialise in a given area, such as toys, household goods, electronics, games, medical products, and the like.

Be sure to seek out a prototype maker with the expertise to design and develop your concepts into the prototype that you envisage. Also, before going into production you need to have considered who's to produce your product, investigated the costs of making any necessary moulds and tooling of the required raw materials, and of packing and shipping the product. An experienced prototype-maker can guide you on these, or indicate contacts who can help.

Meeting product safety standards

You must scrutinise your product for health and safety requirements before introducing it to the market. Check if anyone can get injured by using the product. For example, a buyer reviewing a new stuffed toy may ask, 'Can a baby pull the eyes off and possibly swallow them and choke?' The purpose of health and safety requirements is to prevent personal or property injury or damage.

Developing design control documents

Many products require specifications. Imagine trying to fix a new car or work out how to set the DVD recorder without an owner's manual. If your product demands it, you must create a well-written user's manual. A useful bonus here is that the manual automatically carries copyright – you can sue a competitor who copies it without your permission and you may pocket a nice damages payment. See Chapter 9 for more on copyright.

After you commit your final product design to production, you must document that the design conforms to quality control standards in order to meet the government regulations that apply to your invention, such as electrical, health and safety, and performance codes.

Furthermore, to enable the production to go smoothly, and especially if you intend to produce the product in large volumes, you must document in sufficient detail each and every step of the manufacturing process, with step-by-step assembly procedures.

Considering graphic design and packaging

Packaging plays a major role in attracting the potential buyer to take a closer look and purchase the product. Graphic design imprinted on the box can be eye-catching or dull, depending on the details and colour co-ordinates. Every individual favours a unique favourite colour. However, mass-market appeal, not logic, determines the colour of the final product.

Large companies spend millions conducting research to find out what consumers want to purchase. They know whether 3-year-olds like red better than purple. This knowledge is important when selling millions of products to a specific market. Have you ever wondered why the colour of toy cars for children is more likely to be red than purple or yellow? Market research can tell you. Be sure to conduct a significant amount of research before selecting the final colour for your product. (Chapter 18 discusses how to conduct market research.)

Pricing your product

In order for a product to be successful in the market, it must meet certain *margins*, which is accounting speak for the mark-up between the cost of manufacturing and the price of the product. The margin enables you to pay your bills and have some money left over for reinvestment or pure profit. The nature of the product and where you sell it determines the product's mark-ups.

The minimum mark-up on a product is typically 4:1, meaning that the end purchaser (consumer) needs to pay four times what it costs to make and ship. The minimum mark-up generally arises for products that sell in mass merchandise stores, such as Asda. For example, if the consumer wants to pay no more than £10 for a product, then the manufacturing, packing, and shipping costs for that product can't be more than £2.50 or you lose money.

Mark-up through other outlets tends to be higher, though always remember that product mark-up is what the market can bear and consumers will pay. If a product sells on television, the product mark-ups from the manufacturing company to the consumer may range from 5 or 7 to 1. For example, in order to make a profit on a new televised product that costs £19.99 with a mark-up of 6:1, the manufacturing, packaging, and shipping combined can't exceed

£3.33. The industry in which the product sells and how much people are willing to pay for the product determine the product mark-up. All industries have price points. If you mark up the price too high on the product, say 10 or 20 to 1, most people won't buy it.

Don't make the mistake of calculating the manufacturing, packing, and shipping costs and then multiplying by your given mark-up factor so that you can then tell consumers what they'll pay for your product. People often make the same mistake. You must first ask the potential consumers (through conducting market research) what they're willing to pay for the product, and then divide that amount by your mark-up factor. If your product costs come in over that number, drop the product until you can find a way to produce it for less.

Choosing the right manufacturing materials

The majority of commodity products in today's marketplace are made from plastic materials. Vast arrays of plastic resins are available with various resistances to thermal stress, scuff, compression, expansion, and rough handling. Your prototype maker can help you figure out what type of resin you need. Most of all, each resin requires specific tooling, which greatly influences manufacturing costs. The cost of the resin versus the cost of tooling is nearly always a major consideration.

During recent years, plastic resin manufacturers have become very competitive. They provide a wide selection of available plastic resins to meet the requirements of various products. For example, your home telephone is made of a rugged ABS material with a rubber additive, so it's less brittle if dropped or otherwise abused. This material may cost £2 per kilogram (£1 per pound). The thin plastic boxes your raspberries came in at the supermarket are made of polypropylene, which is a cheaper, more brittle plastic that costs less than £1 per kg (about 40 pence per pound). A reusable hospital device may be made of a thermoplastic material, which can withstand continuous heat sterilisation between uses, costing about £10 per kg (£5 per pound). So when you choose the final material to use for the manufactured version of your product, think about how consumers may use and possibly abuse your product.

Although plastic technology in product function, appeal, and cost factors dominates current manufacturing, many products lend themselves to fabrication from other materials: metal, glass, rubber, leather, wood, chipboard, fibreboard (HDF and so on), paper, or combinations of two or more of these. Many manufacturing plants still have metal stamping, turning, welding, and milling facilities, in addition to woodworking and plastic facilities. Make sure that you select a material that's well suited to your product's intended application and the price the market can bear.

Defining Your Process Parameters

Okay, so your invention isn't a new consumer product but an improved process – a means to operate, produce, or treat in fewer steps, more rapidly, more economically, less dangerously, more efficiently, or with better green credentials. You still need to develop your process to determine the detail of its operating limits and optimum performance.

As for product inventions, you face the question of whether to undertake the development yourself or to get others to help you. If you've the talent and the facilities, by all means do it yourself. But if not, seek out local research and development organisations. And again, a good source of help may be the appropriate department of a local university, be it the department of aeronautics, biology, chemistry, engineering, information technology, medicine, mining, or wherever the field of your invention may lie. Local schools and colleges may also be able to assist. And don't forget the guidance available from local inventor organisations.

You need to define process or treatment parameters that may include such aspects as the required quantities, temperature, pressure, essential catalyst, step sequence, time period, process apparatus, and testing equipment. An experienced research department knows which of these aspects, or others, are likely to be the most critical in the field of your invention.

Progressing Other Types of Inventions

Inventions have a way of not fitting tidily into categories. New compositions of matter – a new toothpaste, floor cleaning liquid, or non-drip paint, for example – may require a different approach from those we talked about in the previous sections. Nevertheless, you may need help to develop and test such inventions, and again, you may find similar support from local specialists. For new software, you may need help, but it's an area where many inventors have the skills and IT equipment to go it alone.

Whatever the nature of your invention, it presents a far more appealing prospect to potential investors, makers, buyers, licensees, or users if they can put it into practice. You can't expect to promote your invention if it's little more than a gleam in your eye. Work on your invention, think about possible variations, and consider whether other commercial applications exist. You may be pleasantly surprised by the substantial improvements your development thoughts and activities achieve.

Setting the Terms for Professional Help

Whatever the nature and field of your invention, get price estimates in writing from the professional helpers (contractors, service providers, and others) for all development and professional services. Ideally, obtain get price quotes from two or three different individuals or organisations so that you can compare them, and find out additional information and suggestions while talking to them. And then make sure that you have written contracts with them to ensure that you own the results of their work for you. If you have existing arrangements with suppliers, make sure that these cover the new services. See below and Chapter 11 for more on setting up contracts with helpers.

Negotiating a contract

Try to obtain a price estimate for the total job, as well as by the hour. When the professional helper charges by the hour alone, their profit depends upon how many hours they spend on your job, which can encourage 'time padding'. Every project is different, so the helper finds it difficult to offer a fixed fee for the job. However, you may be able to agree on an hourly fee with a price cap, to prevent the helper from charging you for too many hours.

When working with a professional helper, always use a signed written service contract (also known as a 'work-for-hire' agreement – see Chapter 11 and the example in Appendix A) as well as a signed confidentiality agreement (see Chapter 2 and Appendix A). The service contract gives you, the inventor, the exclusive legal rights in the results of the professional helper's work. The *confidentiality agreement* requires the professional helper to keep secret your invention, your ideas, and the results of the work.

Many professional helpers show how to improve your product or process, and possibly suggest improvements to it. However, a helper who adds to your invention without a well-defined contract can easily become a co-inventor on your patent. Be careful! You don't want someone who you pay on a short-term basis to become a co-owner of your patent.

The contract must state that you're the owner of the invention and that you hired the helper to do work for you, including improvements, for good and adequate remuneration. If, while making the prototype, the helper contributes to the invention to the extent of being a co-inventor, then do name them alongside the other inventor(s) in your patent. But ensure that the helper signs the contract before the work starts, stating that all intellectual property rights in the project – including patents, designs, trade marks, copyright, and trade secrets (see Parts I and II) – belong to you.

Professional designers, engineers, prototype makers, and development labs are familiar with work contracts. If they create any hassle, save yourself a lot of trouble by walking away and hiring someone else.

During each step of the invention's development, you must have iron-clad contract terms relative to what the helpers in each phase are going to do for how much money, and the date(s) by which the phases need to be completed. If you make changes to the terms after the helpers sign the contract, they may be able to make additional charges for the changes. And remember that although the helpers have responsibility for the quality of their own work, they're not responsible if your idea turns out to be not very good.

Working with an existing manufacturer

Often, a manufacturing company that makes similar products can make a prototype or conduct process experiments for you. The company may not want to license your product or process and pay you a royalty, because it runs its own development programme. However, the company may have slack times when it can work with you on your development and possibly manufacture limited quantities of your product.

For example, bakery workers typically start about 3 a.m. and work until 2 p.m. to maintain fresh products. So, the building and equipment the workers use become available after 2 p.m. until early in the morning. One inventor made an agreement with a bakery owner to use the facilities and equipment during the off time in order to produce special fruitcakes and biscuits for sale through mail-order catalogues.

Protecting Your Ideas During the Development Process

Memories fade over time. Nothing suffices over the life of the invention like a great written document.

Keep detailed and accurate information regarding your invention. For example:

- ✔ **Keep a journal about your invention.** Write down a clear description of your idea.
- ✔ **Keep all receipts as you develop, build, and test your product.** If someone tries to copy your invention and claim it as their own, you need all the detailed documentation to prove your ownership. The legal case with the most convincing documentation and best-kept records wins.

The bottom line is that no one really cares about your invention until someone makes money from it. When you start making money, people come from every direction and claim they helped or assisted you with the invention along the way, wanting to reap financial rewards from your hard work. Define from the outset the exact terms, including the rewards, of working for you must be, and get their written agreement to those terms.

Chapter 11

Hiring Helpers and Using Service Contracts

In This Chapter

▶ Employing professionals to help you

▶ Guarding your intellectual property

▶ Assembling a service contract

*U*nless you're a master of all trades – inventing, engineering, and design-ing, to name just three relevant skills – you need to hire people to help develop various aspects of your invention. In most cases, hiring professionals to assist you in the development process can help you move your invention to market quicker and more efficiently. Many experts exist out there who can work side by side with you.

Of course, you want people with up-to-date experience that's relevant to your project's size and scope. You also want people who are contractually obliged to keep your ideas confidential and respect your intellectual property (IP) rights.

This chapter tells you about various types of people you may need to hire along the way, and discusses the importance of service contracts (also known as work-for-hire agreements) and how and why you must get your model makers, manufacturers, designers, and others to sign the contracts.

Hiring Professionals to Turn Your Idea into a Reality

You have an idea, so now what? To turn your idea into reality you probably need the active help and practical advice of a whole army of people. Forgetting the financial and marketing people for a moment, you may need prototype builders, mechanical designers and engineers, packaging design-ers, software writers, Web designers, and so on to undertake all the practical jobs that relate to turning your idea into a money-making reality.

Some of the types of helper you may need to hire include the following:

- **Bookkeeper or accountant:** Helps with financial matters
- **Graphic artist:** Designs the artwork and wording on your packaging
- **Prototype maker or product designer:** Makes the development product for you
- **Regulatory personnel:** Ensure compliance with government rules and regulations
- **Sales people:** Assist in selling your product
- **Training instructors:** Show people how to use your product
- **Writer:** Puts together an instruction manual that explains how your product works

Make sure that anyone and everyone who gains access to or has knowledge of your invention (its structure, mechanics, components, or any other aspect of it), and anyone who's privy to your financial information, signs a confidentiality agreement (see Chapter 2 for more on confidentiality agreements). Also, ensure that anyone who's involved in any creative work connected with the invention signs a service contract (check out 'Working Out Service Contracts', later in this chapter). Each person you hire carries valuable information and know-how in some fashion. For example, you don't want your bookkeeper (who's not bound by the same professional rules of non-disclosure as an accountant is) to discuss your financial information with his family and friends.

The next sections talk about a couple of the more practical professional types you may need to hire.

Picking a model maker

You have a mechanical or structural invention that you believe has commercial potential, but you're still at the idea stage and don't have a functioning model or *prototype* to determine whether your idea really works. If you don't know how to take your idea and make it into a working model, consider hiring a professional.

A prototype or model maker builds a functioning model of your invention. The prototype reveals any design flaws or engineering difficulties. The process lets you determine the best materials to use and may help you refine your design. When (and if) you get a functioning prototype, you may use that model to work out the specifications for any moulding tools, dies, and stamps you need to produce your invention.

Plus, you benefit from the prototype maker's expertise. A professional prototype builder is apt to be aware of health and safety regulations about the types of materials to use. He also knows about current and changing government standards. Chapter 10 provides details on finding, hiring, and working with a professional prototype maker.

You have to be careful about infringing other inventors' IP rights through your prototype builder. He may bring IP knowledge from other jobs and apply them to your new product. On the one hand, you want him to use his expertise, but on the other hand, you have to be careful of potential patent infringement.

Many great ideas fail in the prototype stage because the prototype, which you create to see whether an idea works, proves that it doesn't. You may pay a pretty penny to develop your worthless prototype, but don't be bitter; you could have pointlessly spent many times the amount you spent on the worthless prototype if you'd continued into even more expensive stages of the development process. And at least you know the truth.

Arranging for packaging personnel

Packaging is an extremely important aspect of your marketing plan. Packaging sells products. We cover the ins and outs of packaging in Chapter 13, but this section talks a little about the people who get your invention into its box, bag, bubble-wrap, or all singing and dancing display unit.

Proper packaging, meaning profitable packaging, requires a variety of skills. You may end up hiring a few packaging professionals. For example, you may need the

- ✔ **Engineer:** Makes sure that your packaging stands up to shipping abuse and stacks easily on shop shelves.
- ✔ **Graphic designer:** Creates an eye-catching visual appeal for your invention.
- ✔ **Sales writer:** Makes sure that the packaging displays the correct, customer-attracting words.
- ✔ **Silent store shopper:** Scans what's currently selling in the shops and studies pricing, packaging, colour schemes, the competition, and any shelf space limitations of retailers' shelves.
- ✔ **Technical writer:** Explains what's in the package and how it works.

Any or all the above mentioned people can claim rights and demand additional compensation for their work unless you bind them with a service contract.

Protecting Your Idea and Your Product

Unless you're an MI5 agent working deep under cover, you may tend to talk about what you do at work to your partner, friends, and even people you've just met. Think how easily such innocent conversations can reveal information that your boss or company wouldn't want people to know. And realise that someone involved with your invention may think that they're making simple conversation at a party, but in reality they're disclosing trade secrets to your competitors.

You need to prevent such conversations in the first place and protect yourself from the damage such talk may do if it happens. The best form of protection is by using confidentiality agreements (see Chapter 2 for more on these). Having employees and contractors sign a confidentiality agreement at least makes them think before they speak, and at worst, offers you some protection should loose lips lead to litigation.

A cautionary tale

A game developer in the US came up with a theme for a fabulous video game. Developing the storyboard and doing the research took months. The developer hired a software writer to develop software that would follow the game theme.

The software writer duly completed the software and the developer paid the writer for his work. The developer took the software to an IP attorney and registered its copyright at the US Patent & Trademark Office (something that's possible in the US but hardly anywhere else – see Chapter 9). The developer thought that he'd done everything properly.

The video game was a tremendous success. The developer licensed the game to one of the largest manufacturers for a 5 per cent royalty, which was the norm at the time, and which translated into millions of dollars in *potential* royalties.

But then the problems came. The developer received a legal letter claiming that he didn't have the rights to license the software. When the lawyers met around the boardroom table with the developer, in came the software writer. The developer and his attorney brought their original copyright registration details – and so did the software writer and his attorney. The developer was confident he held the rights. He didn't. Although both of them had registered their copyright, the software writer had applied and registered the copyright on the software before the developer filed his own copyright application.

You ask, 'How could this happen?' It happened because the software writer completed the software and registered its copyright before revealing the software to the developer. The developer hadn't insisted on a service contract and no evidence existed that ownership of the copyright had passed, or been intended to pass, to the developer. No one could prove otherwise. What was the outcome? With millions of dollars at stake and the market ready, not to mention potential fraud from the licensor towards the licensee (he'd licensed something he didn't own) the royalty fees were split. Both the developer and the software writer each received half the royalties.

In another scenario, say you develop and produce your invention, and your wildest dreams come true – your product sells like hot cakes. Your bank account grows and you're well on your way to easy street when you hear a knock on your door and then one of your hired professional helpers serves you court papers stating that your patent omitted him as a co-inventor. You think, 'What in the world is he talking about?' Then you spot in the court papers the name of the prototype maker who helped you develop your product. You haven't heard from him since you paid his fee, and naive inventor that you are, you thought that was the end of the story. It's not.

You may have sweated blood, borrowed money from your family and friends, begged money from people you don't know, mortgaged your home, and denied yourself (and your family) holidays to turn your idea into a thriving business. Now the nice professional who made your prototype years earlier comes along and wants a piece of your action. To do so, he claims status as a co-owner and wants half of your royalties and half of your sales income. To add insult to injury, under the patent law on joint inventors, the prototype maker may have a good case, despite the fact that you paid him a fair price for his professional services at the time. (See the nearby sidebar 'A cautionary tale'.)

Another potentially troublesome scenario involves the graphic designer who works on the packaging for your product. That nice lady who designed your product's beautiful box may come along and demand that you pay her copyright fees for each unit you sell that's packaged in the box she designed. Unless you secured an agreement to the contrary, the artist owns the copyright to her work. (See Chapter 9 for more on copyright.)

You can prevent such an unhappy picture from becoming your own nightmare by observing the following tips:

- **Negotiate all the terms right at the beginning when you're poor and just starting up and put them in a written, signed agreement.** Money attracts people who want it, and your success may look like a free ride to someone who knew you or your invention in its early stages.

- **Document every step of the process.** Record when and where you met with designers, lawyers, and financial advisers – and what you discussed. If you think of a name for your product, check out whether somebody else already owns it and if not, then seek your own trade mark protection for it. (See Chapter 8 for more on registering trade marks.)

- **Insist on confidentiality agreements.** Ensure that anyone you speak to about your invention signs a confidentiality agreement (also known as a non-disclosure agreement) before you disclose information to that person about your invention. If you want to bounce ideas off a friend or colleague, get them to sign a non-disclosure agreement. Chapter 2 details more on these agreements and how to keep your invention confidential.

✔ **Get people to sign service contracts.** If you want anyone to do any kind of work with your invention that may result in the creation of IP rights, such as the copyright created when they work on drawings, prototypes, packaging, software, or any other aspect, always get each one to sign a service contract along with confidentiality agreements. A signed service contract can stop a lawsuit in its tracks.

Many inventors confronted with legal action over ownership of intellectual property immediately want to blame their attorney, which is probably a natural inclination, but not effective or logical. The attorney's job is to get you a patent or other IP rights, not necessarily to protect all your interests every step of the way. You need to do your own homework and inform yourself of your rights and how to protect them.

Working Out Service Contracts

A service contract (or *work-for-hire agreement)* can specify that you're the inventor who's come up with an idea, and you're hiring the professional services of a person to help your work with your invention. You can require that by signing the contract, the professional helper relinquishes all rights to ownership of your patent application (even if you must name him as a co-inventor under the terms of the Patents Act: See Chapter 5 for more details on contract terms), your registered or unregistered designs, your trade mark, and the copyright in such elements as product packaging or labelling, instruction manuals, sales literature, or operational software. The contract should include a clause assigning (transferring) these rights to you and agreeing to sign any assignment documents required by intellectual property offices to confirm your rights.

Service contracts are typically made between an employer and an employee or between an independent contractor and a person who needs the contractor's professional services to do a specific job.

Employers often require employees to sign a service contract upon starting their employment, and for good reason. Imagine you own a software development company and you hire computer programmers to develop new software to add to your company's product line. You assume the financial risk – it's your money on the line. The programmers have the security of a salary, pension rights, paid holidays, and possibly health insurance. From this perspective, it's only fair that the programmers sign an agreement assigning all IP rights to you, as well as agreeing to respect the confidentiality of your operations and trade secrets both during and after leaving your employ. You don't want your programmers taking your trade secrets and know-how and either passing them to a competitor or starting a competing business of their own.

Some of the key components to consider when negotiating your agreements include the following:

- ✔ **The length of the term of the project contract:** You want to make sure of a beginning and an ending to your project. You want to know when your work is going to be complete. Why? Say that you're working on your prototype and want a working model in order to conduct market research with potential consumers (with, of course, your IP applications or a confidentiality agreement in place). Because you may go through several prototypes in the developmental stages, you need to state that your prototype must be complete in say 60 days, 90 days, or whatever time frame that you and your product developer agree upon. You don't want an open-ended contract. Some developers drag a project on for several months or even years and run up a lot of bills in the meantime. Get your time frame set and move forward.

- ✔ **A statement that says you own all variations and revisions leading up to the completed project:** Your statement must cover concepts that you consider but don't use and prototypes that don't work. Knowing what doesn't work can be almost as valuable as knowing what does. For example, 3M made an adhesive that didn't quite work but went on to use it to create their highly successful Post-it® notes.

- ✔ **Extended coverage to include any subcontractors that the person you hire uses:** You may hire a manufacturer to make the box for your product, and the box manufacturer may subcontract a graphic artist, who's a part-time student, to design the artwork. Often, design and engineering firms subcontract various work to outside firms or individuals in order to curtail employee overhead costs.

- ✔ **The payment schedule, which you base on delivery:** An agreement with a professional helper may stipulate that you pay one-third of his fee up front, one-third when they complete defined parts of the project, and the final third upon completion of the project. Be sure to address how you pay expenses such as travel, material costs, and subcontractor fees.

- ✔ *Scope creep* **– quite often the development process ranges far and wide as you consider and reject idea after idea:** Document your expectations in very clear terms and specifically address what happens in the event of any change of direction or broadening of the scope of the original project. For example, your original concept had two wheels but now you realise that you need four. You must also include the similar situation in which you specified two wheels but the developer used four.

Keep in mind that every project is different, and that the associated agreements can be simple or extremely complex. Your legal advisers may be able to provide you with other elements to consider, based upon your specific situation.

Appendix A provides a sample service contract (work-for-hire agreement) for hiring professional helpers (in this example, for a professional writer of instructional text), which you're free to modify to fit your situation. However, we strongly recommend that you involve your legal advisers. Making the contract watertight and bombproof is generally too important to risk doing it yourself. In commercial deals things can go wrong very quickly. You want to fend off as many potential problems and lawsuits as possible. Seek good legal advice at the outset! (See Chapters 3 and 4 for more on choosing your advisers).

From the beginning, always have everyone sign a confidentiality agreement and always have professional helpers sign a service contract. *Never* leave anything to chance, unless you want to give away half or more of your profits. Negotiate before the idea begins to produce any financial returns and while you're still poor, because when you're successful many people may go to great lengths to get a piece of your profits.

Chapter 12

Evaluating Your Invention's Potential

In This Chapter

▶ Checking viability

▶ Putting your invention to the test

▶ Benefiting from the results

Getting a new product to market takes lots of money and effort from lots of people. So you need to make sure that your invention has a reasonable chance of making it in the marketplace before you expend all that money and effort. Helping you figure out how to determine your product's chances is what this chapter's all about.

Answering Questions about Viability

You travel a long road between waking up with a great idea and seeing your invention on the shop shelves. Each step toward a marketable product is a challenge. You may go down the road of invention any number of times before you hit on an idea that can make it in the marketplace. Rarely is a first invention a marketable invention. As an inventor, you have to be willing to try, try, and try again.

You may need outside help to evaluate whether your idea has what it takes to make it in the cold, cruel marketplace and is worth the time, effort, and money to get it to that point.

Before you get to the multitude of marketing issues, you need to address a handful of basic questions about your invention – questions that you can explore in the following sections.

Is your idea original?

Obviously, if somebody's already come up with, and produced, an item as good as or better than your invention, pursuing a similar idea any further is pointless: You only waste your time and money. Therefore, the most important first step is to find out if your idea is actually original.

You can look in many places to find products similar to yours. If you have an idea for a consumer product, check shops, catalogues, the Internet, and the business and popular press. Conduct a patent search (see Chapter 4 for how to do this). Check trade associations and trade publications in your field and visit trade shows relevant to your idea.

You can't patent your invention if it's in the public domain. Generally, your invention goes into the *public domain* if you disclose it orally or in writing to anyone who isn't bound by a confidentiality agreement. If you disclose an invention before filing a patent application, or without a confidentiality agreement in place between you and the person with whom you disclose the invention, then you can't gain patent protection. Chapter 2 has the low-down on keeping hush about your invention.

How are you to produce your invention?

Many inventors just think of an idea, develop it (at least in their minds) to the point at which they have sufficient detail to support a patent application, apply for their patent, and then assume that licensing it to a large manufacturer is a quick and straightforward process. You probably don't need to be told that this scenario is unlikely. For the small proportion of inventions that do actually make it to market, the process can take three to five years, and require a significant amount of time and money. Inventors need to think through the entire process, including production. You need to consider whether producing the invention is cost effective or even possible.

Taking their ideas to a big company is the first impulse of many innovators. Provide the dazzling idea, they think, and let the giant work out the details. After all, the company has the money, the production capability, and the marketing know-how to make your surefire profit maker succeed. Unfortunately, big companies can be very resistant to receiving ideas from outsiders, they may consider an outside invention a long shot, and may want potential sales of an item to be in the tens of millions of pounds. So gaining the opportunity to show your idea to the *right* person within the *right* organisation can be extremely difficult. Large firms may simply refuse to sign a confidentiality agreement with you before seeing the idea, and refuse to look at your invention until you file a patent application that clearly defines it.

At the other end of the scale, you may be able to produce your invention yourself, manufacturing it at home and selling it by mail order or through a Web site. Of course, if you have the wherewithal to start a company, or already have your own company, you've a tailor-made platform for producing and distributing your invention.

If you design, manufacture, and sell a product yourself, you need to consider liability. If someone is injured while using your invention, you may be liable.

Another option is taking your idea to small- and medium-sized businesses. Many smaller firms are interested in producing quantities far below the threshold of a larger company. But although a small firm may lack the marketing and distribution expertise of a larger firm, you can at least get the ball rolling.

Can your invention actually make money?

Whether your invention can make money is a question designed to keep you awake at night. Unfortunately, no one can answer with any assurance. After all, even major corporations that conduct detailed market research sometimes make wrong decisions.

You certainly can and need to do research to determine whether your invention has a market (see Chapter 18 for details on researching). You need to figure out your profit margins to make sure that you can take in enough money to make production worth your while. If you find that you can't produce your invention cost effectively, don't waste your time. (Of course, even if you find that you can make money from your product, weighing the profit margin after production and distribution costs against your personal efforts, expenses, time, and travel away from home may cause you to think twice before travelling down this adventurous road.)

If you license the manufacture and distribution of your idea to a larger company, your royalty fee is probably less than you expected. To make millions from your idea using the licensing route to market requires the invention to have a very high unit price or – if inexpensive – to sell in the tens or hundreds of millions!

Ultimately, you make your best guess, based on the information and feedback that you gain during the evaluation stage, and either abandon ship or pull out the stops and charge full steam ahead.

Asking for Evaluations

Just because *you* think that your invention is brilliant, it doesn't guarantee that everyone else does. And because you won't make money selling your product to yourself, you need to know what other people think about your invention. Solicit their opinions before you spend a lot of money developing your invention. The best people to ask are anybody and everybody you can convince to talk to you.

If you haven't already protected your invention with a patent or registered design, make sure that you have a confidentiality agreement in place before disclosing its specific details.

Hiring a professional evaluator

Commissioning a professional, unbiased product evaluation before committing yourself to the patent application process is a wise move. You may not want to spend time and money preparing a patent application unless you can show that your invention has genuine commercial potential.

Take care when hiring an evaluator – a number of invention promotion companies operate nothing more than scams. See Chapter 19 for more on these companies.

Getting some free advice

A professional product evaluation service is a good way to receive impartial, constructive feedback on your idea, but even people who aren't professionals in commercialising new products can be an excellent source of information for you. You may consider speaking to the following:

✔ **Buyers and salespeople who buy and sell products similar to yours:** They work with the public every day and have a vast amount of experience. The manager of a large retail department store is likely to have ten or more years' experience in the business and be familiar with pricing, warehousing, distribution, packaging, and more.

✔ **Engineers and technical people:** They can give you advice regarding product design.

✔ **Entrepreneurs:** They can tell you what running a business is like, give you tips on what

to do right, share what they did wrong, and offer advice on what they'd do differently.

✔ **Packaging designers and companies:** Packaging sells products and these people can maximise your product's appeal to its target market.

✔ **Professors and students studying your industry:** Teachers and students can give you additional insight into the current market and they probably have information on the latest research.

If you don't know people in the above areas on a personal basis, find someone who does. You don't know the consumers who are going to buy your product, but you have to figure out how to get to them. Plan and target who you want to talk to and then locate and contact them.

You also need to protect yourself against inexperienced and incompetent evaluation services. Some organisations, including universities and non-profit associations, use undergraduate students and other untrained or inexperienced staff as new product evaluators. The evaluation system the company uses may be great, but the experience of the evaluator or evaluation team is what counts. Trained and experienced product evaluators can spot technical and commercial flaws that others can't.

Make sure that you know exactly what the evaluation includes. For example, some companies perform a comprehensive patent search and include the results and copies of conflicting patents as part of their service – this extra information can save you additional time and money.

Ask the company how many products it gives a high rating to in comparison to the products that get low ratings, and don't be put off if the company gives you a report that points out a lot of flaws in your product. You pay the company to do exactly that. Finding out the negatives before you spend money on getting patents or on producing your invention can save you big bucks, whether you decide to proceed or not.

Ways of finding out how to have your invention evaluated include

- Business Eye, Wales. `www.businesseye.org.uk`.
- Business Gateway, Scotland. `www.bgateway.com`.
- Business Link, England. A useful resource for impartial, practical advice. Visit `www.businesslink.gov.uk`.
- Highlands & Islands Enterprise, Scotland. `www.hiebusiness.co.uk`.
- InvestNI, Northern Ireland. `www.investni.com`.
- Ideas21. A London-based innovation network that is devoted to the successful exploitation of inventions and intellectual property. Visit `www.ideas21.co.uk` for more information.
- The British Library Business and IP Centre. Packed full of useful information and where you can also attend enterprise workshops. Visit in person or go to `www.bl.uk`.
- Trevor Baylis Brands. The company set up by well-known inventor, Trevor Baylis to help other inventors with their ideas. Web site `www.trevorbaylisbrands.com`.

Bracing for feedback

As the inventor, you already know your invention's good points (though you won't mind evaluators repeating them). What you need from evaluators, however, is a clear idea of whether your invention is worth pursuing down the long and costly road to the marketplace. You want solid information that helps you avoid costly mistakes.

Look for and listen to suggestions for improvement and constructive criticism. Making changes before you go into production is much cheaper than making them afterwards.

A comprehensive professional evaluation gives you:

  ✔ An overall assessment of your invention's commercial viability

  ✔ A summary of the intellectual property situation – whether the idea had protection before or has it now, and in which territories any protection is in place

  ✔ A heads-up about problems that may arise as your product moves through the different commercialisation stages

  ✔ An indication of whether a prototype is necessary and useful to present your invention to others

  ✔ A pathway to determining what additional information you need in order to further investigate your invention's overall feasibility and marketability

An honest, unbiased product evaluation report and documentation of any action you plan in response to it can help you attract investors, potential licensees, and others who can work with you in bringing your product to market.

Rip-off or 'scam' invention marketing companies are able to snatch inventors' money because people *want* to hear that their products are winners and that they'll make millions of pounds. As a reality check, ask everyone who tells you that your idea is worth millions to invest in it! After all, you wouldn't want to deny them the opportunity to invest in a multimillion-pound deal, would you? If they balk, you know what they really think.

Looking at evaluation techniques

When people evaluate your idea for viability, you may get one of several types of response:

  ✔ **Simple 'yes, your idea is great' or 'no, your idea stinks' evaluations from one or more evaluators:** This type of appraisal is simple but useless. Even if you just ask your friends, ask them *why* they think that your invention has a shot or not.

✔ **A comprehensive, individual analysis by one or more specialists who have technical and/or marketing expertise in your invention's area:** This is an ideal evaluation, but may cost several hundred to several thousand pounds. Think very carefully before spending such large amounts of money and find out exactly what you get and from whom.

✔ **A systematic analysis by a cross-section of people who represent a broad range of technical and commercial expertise.** This type of evaluation is just right. It gives you the level of detail you need to make decisions about your invention, and the evaluation costs are within reason (usually around £300).

Any evaluation system you use needs, at a minimum, to accomplish the following:

✔ Identify ideas and inventions worthy of further development

✔ Recognise ideas and inventions that don't have the potential to become successful innovations

✔ Point out potential trouble spots or areas that require special attention before development or commercialisation begins

✔ Suggest strategies for further development and/or commercialisation

Predicting whether a product has what it takes to succeed while you're still in the early stages of the innovation process is hard to do. Too many unknowns exist for even a professional to predict success accurately. Very rarely, an idea or invention has such clarity of technical and commercial feasibility that its potential for success is obvious. However, most ideas and inventions require astute development and marketing in order to become successful products. A professional evaluator won't know how you plan to market your invention, what it may end up looking like, or any of a hundred other key factors that contribute to your product's success or failure. An evaluation can help you make the decision about whether to go ahead with development, but it can't guarantee or even predict whether your invention can make money.

Assessing all aspects

You need to evaluate your invention from a variety of perspectives. The following sections outline some areas to consider.

Health, safety, and welfare

If your invention is subject to government regulations of any kind, you need to be sure you can meet them. Think through safety issues by considering the potential hazards of using your invention.

Take environmental impact into account. Does manufacturing your invention require scarce or fragile resources? How eco-friendly is the manufacturing and disposal process? Can your invention promote destruction or misuse of natural resources?

Ask who your invention benefits. You want to make a positive contribution to society as well as to your bank balance.

You also want to take care of your own welfare by determining whether you can protect your invention with a patent, registered design, trademark, or copyright. See Chapter 3 for more information.

Development, feasibility, and function

Evaluating what stage your invention is at helps determine what more you have to do. If you're just at the idea phase and things don't look good in other evaluation categories, you can cut your losses or switch directions.

A reality check is also necessary. Your idea may be ground breaking, but if it doesn't work, isn't practical, or you can't produce or replicate it, you won't get very far.

Beyond the theoretical questions, you need to get a firm grasp on exactly what your product is, what it does, and what purpose it serves. Does your product fill a psychological or physical need for the consumer, or is it a gimmick?

Consider the potential for adding complementary products or offering a range of styles, qualities, and price ranges. Some products can spin off other ones. For example, you could develop a new boating life jacket that can also be used in aircraft. Then you may want to add additional products to your product line for both the boating and aircraft industries, because you'll talk to the same buyers and industry reps.

Look at your invention and compare it with existing products in terms of attractiveness, durability, price, and value. Judge whether your product is compatible with existing consumer attitudes, methods, and uses. Can potential consumers readily recognise its function and usefulness?

If your product requires frequent servicing and parts, estimate the cost and determine who pays them – the manufacturer or the consumer?

Costs

Consider costs throughout the whole process. Get an idea of how much it's going to cost to get your invention to the production stage, including the research and development aspects, the market analysis, legal costs, and so on. Beyond production, you need to consider distribution costs and account for potential problems you may encounter when establishing distribution channels. Also think about marketing: How many resources and how much money and time is it going to take to promote the advantages, special features, and benefits of your product?

Think about whether your invention needs investors to bring it to fruition and if so, who and how much money they may invest. Think about how you (and your investors) can make money from your invention. You need to get an idea of how and when your invention will make enough to pay potential investors off and bring you some profit.

Market matters

You need to determine what kind of market research you need in order to investigate the commercial viability of your invention. The search tells you what the overall market is and the potential sales volume. You want estimates in terms of local, regional, or national sales.

Look at the existing competition for your invention. Is your invention distinctive enough to break into the market? List the advantages your idea has over similar products and look for ways to offset any disadvantages. Even if your product has a shot at taking on the competition, you need to anticipate the appearance of competing products and evaluate how well you can do against them.

Identify barriers to entering the market. For example, does your invention require that you educate consumers about what it does or how it works? Is your invention a supplement to something else, and if so, what happens if that primary product takes a nose dive?

Gauge how stable the demand for your invention is likely to be and whether it's likely to remain constant, grow, or decline. Try to anticipate seasonal or other market fluctuations. You need to calculate your invention's likely life cycle. How long can you profitably sell your product before the market shifts or a new product or technology replaces your invention? Check out Chapter 18 to find out more about researching the market for your invention.

Management issues

Judging how much technological, financial, and general management talent your product requires is an important evaluation step. Knowing who the players are and how they can help to take your product to market is key to improving your chances of success.

If your product requires a significant amount of technical expertise, having access to that expertise is helpful. Likewise, a knowledgeable financial person on your team is an asset when calculating cost issues.

Risk factors

Consider competition, difficulty of distribution, cost of reaching your target market, financial investment, and technical needs to determine how risky your venture is. Starting any business or introducing any product to market is risky, but you need to assess whether the anticipated return for your investment in time, energy, and money is worth that risk.

Benefiting from the Results of Your Evaluation

Product evaluation shows that your idea is worth spending time, money, and energy on. Use the evaluation to the fullest extent by showing it to potential investors, the business adviser at your bank, and your accountant. Remember, these people probably know people who have money to invest.

The results of your evaluation may reveal a short product life cycle. A short life doesn't necessarily mean a profitless life, however. Fad items can make their inventors a fortune, if they use certain financial, product, and marketing strategies to maximise their profits. In this scenario, the evaluation system can be a useful mechanism to help avoid obvious mistakes. Unfortunately, even the most sophisticated marketing-orientated companies make mistakes that they may have avoided had they conducted a more comprehensive initial evaluation of a new product. Don't fall into the same trap.

Chapter 13

Going into Production

*H*ow many would-be inventors progress beyond the dream of product conception to the reality of a manufactured and fully functioning product? Fewer than you may think. The few inventors who persevere and develop their invention into a functional product are the true entrepreneurs. If you want to be one of the successful ones, you must avoid the countless traps set to trip you up. You have to be or become an inventor with hands-on, practical knowledge of what it takes to be a manufacturer.

This chapter outlines the manufacturing issues you need to be aware of, whether you make the product yourself or subcontract the manufacturing process. (For details on subcontracting, see Chapter 17.)

Focusing on the Process

The production process starts with you – you conceive the product and possibly patent it. After that, someone else (usually) designs your product, the government may inspect it or regulate it, and, finally, a manufacturing plant produces it.

This section explains the process and how it works.

Whether you subcontract the manufacturing or make the product yourself, you need to know what producing your invention involves. You must be able to discuss manufacturing components and understand the costs of the production process so that you can make intelligent decisions.

Examining product cycling

Everything in life has a beginning and an end. Sometimes those starts and finishes are hard to see, but you don't have that fuzziness in the production process – you start with a collection of raw materials that don't add up to anything and end up with something you can see, touch, use and, most importantly, sell.

Product cycling refers to how fast you can make, ship, and sell your product before a re-order comes in. It is also called *turnaround time*. Product cycling requires you to properly address all contributing factors before the cycle starts. A tremendous amount of preparation is necessary before an assembly line can start producing something as simple as a key ring or as complex as a car. One thing out of place – raw materials aren't up to standard, or there aren't enough workers to run the production line, for example – and the whole process breaks down. Planning is crucial for any endeavour, and for a production operation, planning is everything.

The greater the number of diverse components in the final assembly process, the more critical long-range planning becomes. Manufacturing processes can be extremely varied depending upon your product and its level of sophistication. Today, you have a wide variety of plastics, resins, metals, and manufacture tooling to deal with in order to make even a simple product.

For example, say you develop a new toy that uses a special lead acid battery (a miniature of your car battery) and several different light bulbs. Your toy requires injection-moulded plastic parts, metal parts that need to be turned on a lathe, and some stamped metal parts. You also need to make and assemble the special battery, the light bulbs, and maybe a battery charger. And you need a number of nuts and bolts to hold all the parts together. On top of keeping track of all the components, different companies in different places may make different components. So you have to co-ordinate getting all the materials to the right manufacturer at the right time, and then everything assembled at the right place. You can see how production scheduling can cause ulcers.

Your ulcers only get worse because your invention is a toy, which has a definite high-selling season – before Christmas. You may plan to make 12,000 toys the first year in 12 production runs of 1,000 each. However, to accommodate Christmas sales, you need to deliver 5,000 units to your wholesalers by – but not before – the middle of August. So you have to work out how to schedule your production runs to meet your Christmas deadline, without needing to warehouse your toys before you ship them. (Retailers won't accept merchandise too far in advance – they have space management issues of their own.)

Considering subcontracting

Most amateur inventors who want to mass-produce an invention subcontract the actual manufacturing. Unless you're already in the production business, you're far better off farming out production of your invention. If you're not already knowledgeable about the ins and outs of manufacturing, we advise you to pay an experienced manufacturer (or two or three) instead of trying to master a very complex process in what's bound to be too short an amount of time.

Whether you subcontract production or do it yourself, it pays to have a general knowledge of the production process for your invention. To start out, you have to evaluate candidates for one or more aspects of the production cycle, and then you have to troubleshoot snags in the process at some point, and you must evaluate your contracts with your suppliers, producers, and distribution vendors periodically. So it pays to know what you're talking about with all these people.

When shopping around for a subcontractor, first consider the ability of the management to run a tight, efficient operation. Some of the issues you need to address are:

- ✔ **Financial stability:** A production plant with huge loans outstanding may have poor management or be on the verge of repossession. Get references from banks, if possible, and from long-term customers.

  Be wary of a bid that's significantly lower than other bids for the same job. The manufacturer may have poor management or be so strapped for cash that it's looking to get jobs any way it can. The company won't be able to continue to offer the low price or won't be able to pay its own bills, and may very well leave you in the lurch.

- ✔ **Overall capabilities:** How efficient is the manufacturing process? How quick? How many components of your product can one plant handle? (Generally, the more components made at one place, the more cost-efficient the process is for you.) Is there room in the schedule to do additional runs if your product sells out quickly? (Work out contingency plans to cover such a situation – and the reverse – in the contract you sign.)

  You don't want to have worked tirelessly for years to conceive and develop your invention only to discover that you can't meet the required sales volumes, because your subcontractor doesn't have enough capacity.

- ✔ **Ability to meet standards:** Your subcontractor must meet not only your quality standards, but perhaps government rules and regulations as well. Make sure that the plant manager is aware of, and familiar with, any regulations that apply to your product.

The manufacturer knows how to set up an efficient mass production line. The manufacturer designs the production line based upon the volume, whether short or long runs. The cost of setting up and breaking down an assembly line between product runs can have a major impact on the final cost of your product.

Remember that a number of manufacturers that make seasonal products take on other products during their slower production times. For example, a jet-ski manufacturer may also manufacture snowmobiles for another company during the off-season.

This time it's personal! Make sure that you visit a manufacturing company if you're thinking of signing on the dotted line for that company to make your product, even if the plant is overseas. You'd be amazed at the number of inventors who don't make this simple check. Especially with offshore production facilities, you need to be confident that the facilities reach your standards. Many inventors have paid foreign manufacturers in advance for 10,000 units or more, only to find that those units were not at all up to the standards promised. And trying to get your money back in a situation like that is a story in itself. Always visit the companies that you plan to do business with.

Chapter 17 covers everything you need to know about subcontracting with a manufacturer.

Making Up the Materials

Production tooling and raw materials form a major part and a major expense of the manufacturing process. Working out where to get your raw materials and how much you have to pay is an important step in the production process.

Working out what you need

Before approaching any manufacturer, you need to have a production-quality prototype. See Chapter 10 for more about prototyping. You also need a *CAD* (computer-aided design) drawing of your product. A CAD is a mechanical drawing that you construct on the computer rather than on the old-fashioned drawing board. Often, if your funds are limited you can conduct research with consumers to see whether a market exists by using a 3D CAD drawing, rather than spending money on a prototype. The prototype and drawing together detail each individual part that makes up your product. Each component drawing is complete to the point where a toolmaker can figure how much it costs to make the component in various materials so that he can bid on tooling each component part, basing costs upon the expected level of volume. (Your projected sales volume figures come from your market research; see Chapter 18.)

You use different materials for different purposes. For example, an aluminium production tool may make 20,000 units; whereas a hardened steel tool, costing a few thousand pounds more, may make 200,000 units before wearing out. So whether you anticipate relatively few sales of a high-priced item or high-volume sales of an inexpensive product helps determine what you need to spend on manufacture tooling. You can develop a cost for each component as part of the overall tooling costs.

If you do short production runs, you have to account for the time required to set up the assembly line. You spread the costs of setting up the line over the number of units that a specific run produces. Remember that you need to store the unique components of each set-up during their inactivity. Consider how and where you can store the components to maintain a state of readiness.

Some of the factors to consider when determining the production facilities and equipment that you need include:

- **Cost:** How can you get the quality you need at the cheapest cost? (See the section 'Calculating Costs', later in this chapter, for more on costs.)
- **Material:** What type of material – plastic, metal, wood – or combination does your invention require?
- **Quantity:** How much of the raw material do you need to have in stock at all times, and where can you store it?
- **Source:** Where can you get the raw material?
- **Storage:** Does the storage involve special requirements? For example, chemicals and resins, food-related products, and so on, may need certain storage conditions.

You may consider listing a variety of sources, thus giving yourself price estimates for the cheapest cost with the best production. In other words, create options for yourself.

Taking on tooling

Look for a manufacturer that makes a product similar to yours. That way, the subcontracting manufacturer has people skilled in developing the tooling needed to produce your invention.

The overall efficiency of any production line directly relates to the design of the tooling required to make the components. The tooling determines the quality of parts that come off the production line.

Finally, the more efficient the design and tooling of your product, the cheaper it is to make your product. Most manufacturers have an in-house packaging division due to cost efficiency. Because manufacturing is so specialised nowadays, sometimes if you're a start-up or small company it's less expensive to get one company to make your product and then another to package it.

Controlling inventory

You aim to achieve a rapid turnover on your inventory because the less money you tie up in raw materials and finished goods, the more money you have to reinvest in your business, or take out in profits. Good inventory control means balancing goods and materials coming in with goods and materials going out so that you have neither shortage nor over-abundance.

The main purpose of installing and keeping on top of a good inventory control system is to save money. Everything costs money, especially in the business world. If you receive the raw material to make your product before the production line is ready to use it, you must spend money storing the material. If you receive the material too late, lost production time, missed schedules, and so on, cost you money.

Packaging for profit and protection

You must evaluate the packaging of a potentially breakable product relative to the type of shipping container, which is designed to comply with the various regulations of the shipping service. When you send a parcel through Royal Mail, you have to follow its rules on packaging and size. The same goes with your product's packaging when you send it by train, lorry, ship, or postal carrier. All transportation types have specific packaging requirements, so be sure to check before attempting to send several thousand units to a warehouse or shop.

How many times has something you ordered from a catalogue or online arrived in pieces? You may be inclined to blame the delivery service instead of placing blame where it belongs: with the person who designed the packaging. In reality, just about everything you use is breakable. Just give your invention to a 12-year-old kid and discover the many ways he can break it. Say you make high-tech electronic equipment: You want to make sure that the equipment is safely packed, most likely with custom foam protection around the product so that it can't move and break in transit.

Packaging has to work on a number of levels and serve a variety of purposes. It has to be sturdy yet attractive, stackable yet easily separated, and able to take lots of handling. And your packaging must make people want to pick up your invention.

The cost of packaging can easily exceed the cost of the product it's designed to protect. For example, a large, fragile, thin-walled glass vase made in Timbuktu at a cost of 20 pence has to travel by mule to a distant harbour and then by ship to the UK, where a lorry picks it up, delivers it to a warehouse, and later transports it from the warehouse to the retailer's shelves. Obviously, the cost of packaging the vase to protect it adequately whilst in transit far exceeds the cost of the vase itself.

Keep in mind that you must package well, but packaging costs can literally wipe out your profit margin if they're too high. For instance, if you have a product that retails for 99 pence and the cost for packaging the product is 79 pence, you've got a problem. To avoid spending too much, get estimates from various packing companies. You can find such companies on the Internet; just type 'packaging companies' into a search engine.

Sorting out shipping

Shipping containers serve two functions: They contain your product and protect it. Containing is important, but protecting is paramount. Consider the way you treat shipping containers yourself. If you've ever received a box at home or at work, you may have dropped it, sat on it, stepped on it, rolled it, kicked it, shook it, and generally abused it in every possible way. When you design shipping containers for your invention, keep people like yourself in mind.

Shipping crates can also be exposed to extremes in temperatures and weather conditions, so you have to make sure that the container can withstand heat, cold, humidity, rain, sleet, and snow.

How would you like to be in a small aeroplane and have a large bottle of exotic perfume break in the cargo hold and fumigate the whole aircraft? Bad packaging can affect many lives.

The following list describes the most common types of packaging. Remember, your product may use one or more of these options at different stages of delivery:

- **Corrugated cardboard box with cardboard separators:** Packagers use this for the bulk shipping of wine bottles, as well as other multipack items.

- **Hard plastic shell that supports and seals in the product:** This works well for computer leads, but not for breakable glassware.

- **Liquid foam packaging:** The packager sprays a liquid expanding foam into a container, and then covers the foam with a plastic paper sheet. The packager then pushes the product down into the container. Finally, the packager uses another plastic sheet to cover the device, followed by a second application of liquid foam over the top. At this point, the packager closes and staples the corrugated box. The foam continues to expand, firmly encapsulating the product within the container. Liquid foam packaging is probably the most expensive but effective method of shipping any product, usually reserved for expensive, potentially breakable products.

✔ **Pallets:** You use these for heavy products. The packager straps the product to a wooden base plate (*pallet*), designed to be handled with a forklift truck. Often, a cardboard box covers the device, which the packager positions by stapling the base pallet. People often use pallets to stack and ship a number of square boxed items to a warehouse.

✔ **Plastic air bags:** You use these to protect and separate mixed items that are pre-packaged.

✔ **Plastic bubble wrap:** This is very convenient and effective but quite expensive.

✔ **Preformed styrofoam:** This encapsulates the product and is designed to snuggly fit the corrugated container and/or other containment box.

✔ **20-foot or 40-foot containers:** Companies generally use these for large bulk shipments from overseas.

Some commercial delivery services may require you to package your product in a specific way. FedEx, UPS, DHL, and other carriers each have specific packaging requirements. Make sure that your product conforms before spending lots of money on packaging. For some products, such as high-quality medical or electronics products, you may need to inject foam packaging around them. Otherwise, the products won't pass inspection by the commercial carriers and you may not be able to insure them. Having a safe type of packaging for your product to prevent damage during the shipping process is definitely to your advantage.

Shipping a new product from the manufacturer to the retailer is a major cost. For example, if a manufacturer in China makes your product, you may not be able to afford airfreight if you have a bulky, heavy product with a low cost value. Sea freight is a better option as it's much lower in cost; however, time can also be a major consideration, especially if your product involves perishable goods. After arriving at the docks or airport, your product then requires shipment to a warehouse by ground transport: road or rail.

Delivering the goods

Where your production plant is physically located in relation to where the customer is located can often mean the difference between the success or failure of your company. The cost of domestic transport may equal or exceed the total cost of the delivery, even from overseas. Costs vary according to whether you deliver to a warehouse or to a retailer. Obviously, direct deliveries to the retailers' warehouses save you considerable handling and transportation costs.

Storing things away

Warehousing can be a major expense. Consider the following costs:

- ✔ Fire, damage, and theft insurance
- ✔ Loading, unloading, and reloading
- ✔ The space itself
- ✔ Transporting goods or materials to your warehouse

If your product has special storage needs (for instance, it requires cool temperatures), you have to accommodate and pay for those as well.

About now, the option of licensing your product for a royalty may look very appealing when you start to understand the real cost and effort it takes to move your product from prototype stage to the shop shelf.

Inspecting Facilities

Finding a space in which to produce your product is a complicated and challenging process. If you have your own manufacturing plant, you have questions on top of questions to consider, but even if you're subcontracting the production of your invention only, you've plenty to think about. The following sections discuss buildings and equipment.

Finding housing

Sometimes inventors visiting a huge manufacturing plant are unduly impressed with the facility itself. Remember that bigger is not necessarily better and that the size of the factory has nothing to do with the overall cost- effectiveness of your product. Basing your decision about a subcontractor on the physical space is a mistake. What's inside the building is what really counts.

A huge building may be impressive to look at, but the company may be deeply in debt trying to pay for the space. The more money the manufacturer owes, the more interest it has to pay and the less competitive it is.

At the other extreme, beware of a bare room. To facilitate quality work, the manufacturer must equip each production line table, or *work bench*, with overhead lighting as well as drop-down utilities, such as electrical services, compressed air, inspection devices, and so on.

Neither too big, nor too small; you want your production facilities to be just right, and have potential for expansion if your invention takes off and warrants increasing production runs or expanding the plant.

Other issues to take into account when you search for a facility include:

- **Shipping facilities:** You need to get materials and possibly partially assembled components to the manufacturing facility. You also need to get the finished product to its next destination. The plant needs to be able to receive and send your product safely and efficiently.

- **Storage space**: You probably need to store raw materials, or finished products, or both. Make sure that the space you need, in the condition you need, is available at the plant itself or nearby.

If you decide to buy production space, you have a few additional issues to concern yourself with. If you intend to manufacture and sell, you need to find out about local planning and other regulations, and you have to consider space for parking and facilities for employees – including toilets and break rooms.

Lining up your equipment

The equipment you need to produce your invention comes in all shapes and sizes; unfortunately, it also all comes with a pretty expensive price tag. But you need the equipment that's best for your invention and even if you can skimp on equipment, doing so isn't worth it in the long run.

If you subcontract the production of your invention, ask the plant manager about the equipment available and make sure that the factory has what you need. If the plant has to install new machines in order to manufacture your product, be very careful. Make sure that the workers know how to handle the equipment efficiently and correctly – you don't want people experimenting on your baby. Find out how fast, adaptable, and safe the equipment is, and whether it can handle the type of material your product needs.

If you do your own production work, you need to determine whether you want to buy or lease your equipment. Unless you already own a factory with the right equipment, we recommend that you lease equipment. A lease provides you the use of equipment for specific periods of time at fixed rental payments. The vast majority of businesses go down the leasing route, because it makes the most sense for novices. The benefits of leasing equipment are numerous:

- **Leasing is flexible:** Companies have different needs, different cash flow patterns, and different and sometimes irregular streams of income. Therefore, your business conditions – cash flow, specific equipment needs, and tax situation – may help define the terms of your lease. Leasing allows you to be more flexible in the management of your equipment.

✔ **Leasing is cost-effective:** Equipment is costly. If you own your equipment, you most likely pay for it over time and can't afford to buy the newest equipment. Furthermore, instead of having to account for depreciation you can simply deduct the lease payment as a business expense.

✔ **Leasing helps conserve your operating capital:** Leasing gives you financial flexibility while keeping your lines of credit open. You can avoid large cash-down payments.

Although leasing does provide benefits to business owners, you have to be aware of hidden costs. Disadvantages of leasing may include:

✔ **Agreements that tie you in:** When entering into a lease contract, you agree to make all the lease payments to the end of the term. You can pay off the lease early, but even if your invention doesn't fly in the marketplace and you stop using the equipment, you have to pay off the lease. In some cases, though, you can sublease your equipment. Read the fine print in your lease.

✔ **Insurance:** A lease agreement may require you to insure the equipment against fire, theft, flood, and so on. If you own or are buying the building that houses the leased equipment, the insurance on your building should cover the contents as well, but you need to make sure. Depending on the type of equipment, you may need additional insurance.

Processing plastics

You can manufacture plastic components using a variety of techniques. Injection moulding can make parts almost as precisely as those machined out of metal. Blow moulding, on the other hand, is quite inexpensive – perfect for parts that don't have to be precise, such as milk jugs.

The skill of the toolmaker and accuracy of the production process determines the quality of the manufactured component. Moulding processes include:

✔ **Blow moulding:** The machine forces a hot glob of melted resin into the mould-tool that's the shape of the desired part. The air pressure within the glob of melted resin forces the resin out around the tool, creating the component.

✔ **Injection moulding:** The machine melts small beads of resin or plastic in a heated cylinder at high temperature. When the cylinder is full, it forces out the resin and injects it into the mould-tool cavity. The machine then cools the mould rapidly and the material hardens. You can use different types of mould tools to get a variety of products – like using a set of biscuit cutters to make a variety of different shaped biscuits.

✔ **Vacuum moulding:** The machine heats a plastic sheet, which is draped over the top of the component mould. A vacuum is applied which sucks the soft sheet over (or into) the mould to form the component. The machine then ejects the part from the press and cuts off the excess material.

Read the fine print to know what your policy does and doesn't cover. And consult with your insurance provider to make sure that you understand everything thoroughly.

Calculating Costs

In order to get a complete picture of your overall business costs, you have to know how much your production costs are. Measure the costs of your day-to-day operations to ensure that things like processing or production downtime don't eat away your profits. In order to determine the final cost of your product, you have to calculate the fixed costs and variable costs on a per-unit basis.

- ✔ **Fixed costs:** Prices that remain relatively stable over time, such as rent and utilities. Even if your variable costs increase, your fixed costs remain the same, or close to the same.
- ✔ **Variable costs:** Prices that fluctuate according to usage or market changes. The cost of materials and labour are examples of variable costs.

Periodic production reports allow you to keep your finger on potential drains on your profits and also provide feedback on your overhead expenses.

Don't get carried away and imagine that your invention can immediately put you on easy street. If you don't carefully plan your production costs, manufacturing your product may just as well put you in the poor house.

Without understanding the costs, you definitely won't understand whether or not you're making a profit. Understanding your ongoing operational costs on a daily basis can ultimately determine success or failure, regardless of how good your product is.

Finding ways to save

One way or another you pay for every aspect of the production process, from raw materials through the production run to transportation. You work out your operating costs based on how much each unit costs to make. You want to first negotiate the best deal up front – the lowest price per production run. Other ways to save include:

- ✔ **Order raw materials and components in bulk:** If you order and receive shipment on mass quantities of resins or nuts and bolts, you can negotiate a substantial discount. Of course, you need to make sure that storing the material doesn't cost more than the amount you save. You also need to be confident that you'll use the entire order.

> To get a feel for how volume can affect price, consider sending a quotation request out to five suppliers in different geographical areas for different quantities of materials. You'll be amazed at the difference in costs between different vendors, as well as the decrease in unit prices as batch volume increases.
>
> ✔ **Buy early:** Many manufacturers offer discounts or dated billing as incentives to buy early. This is important to manufacturers because of production planning and the lead-time necessary for ordering raw materials. If suppliers know that they can count on your order, they reward you for giving them peace of mind. You may end up paying less than usual, or paying later than usual.

If you place a last-minute order, be prepared to pay a premium. The manufacturer charges the additional worker and equipment time to you.

Most businesses operate according to the just-in-time principle. The company schedules supplies, goods, and even employees to arrive right when it needs them and not a moment before. Production facilities make use of just-in-time to form a seamless on-time assembly process. The lack of one component part can cause an expensive interruption of the assembly process, as well as the loss of a time-sensitive order.

Judging economies of scale

Economies of scale basically means that as volume increases, the cost per unit goes down. So, in producing your invention as well as in selling it, you want to maximise production and sales volume in order to minimise costs.

Here's an illustration of economies of scale. You're freezing cherries, and to prepare the cherries for freezing, you have to remove the stones. To do so, you have a number of options: removing the stones one at a time with a kitchen knife, purchasing a bespoke hand-tool that de-stones each cherry individually, or purchasing a cherry processor that de-stones 30 pounds of cherries in one go.

If you plan to freeze only a small amount of cherries, it makes sense to spend a little extra time and use the kitchen knife. If you're freezing several pounds of cherries, buying the bespoke hand-tool is probably worthwhile. If you're supplying cherries for a bakery that makes dozens of cherry pies each week, you have to process hundreds of pounds of cherries. You now must rationalise the cost of paying workers to remove the cherry stones, which becomes a consideration. You can invest the hundreds of pounds required to purchase the automatic cherry stoner because of the high volume of processing required during each seasonal processing operation.

You also consider time, which as we all know, is money. If you're supplying the bakery, the volume justifies the increased cost of buying the tools. In the end, the tooling investment decreases the cost of removing each cherry stone. Another consideration is the time it takes to set up your cherry-stoning production line. After you set up the line, though, you don't incur that cost any more and each cherry costs less to de-stone.

Working with People

Consider the type of labour skills needed to run and maintain the equipment you use. A frequently overlooked critical consideration is the quality and the know-how of production line employees.

You may want to take the time and list the specific types of workers you need individually, as well as the number of people you need. Then work out the average wage rate plus benefits – and gasp at the cost of skilled labour.

You need to be sure that you can hire enough of the specific type of manufacturing workers you need – including machinists, assemblers, packing people, sheet metal workers, metal turners, metal stampers, blow moulders, vacuum moulders, injection moulders, quality control agents, and so on. Do your homework and find out what you really need.

Access to the skilled labour you need is crucial to the success of your venture. When large businesses consider relocating, they take the time to fully investigate the availability of workers. You may want to contact local recruitment agencies for advice. Ask around until you find the answers to help you make educated, cost-effective decisions.

You also need to consider the indirect labour you need to keep your plant up and running – grounds staff, cleaning staff, and so on. Think about how many people you need, their skills requirements, the availability of the labour, and the average wage rate for these individuals.

Checking Quality Control

Consumers have many options when purchasing new products. If your product is poorly made, you lose the customers you have and don't gain new ones. An important rule of merchandising is: 'It doesn't matter what you say about your product, it's what others say.' You need to check quality at every step of the process to ensure that you make your invention the best it can be.

Good quality-control checks ensure that the product gets manufactured in the most cost-effective way and is of the finest quality it can be. To ensure quality you need high-quality raw materials, skilled workers, and a smooth assembly process that manufactures your product without flaws the first time: no bubbles, no cracks, no tears, no speckles, no sharp edges, no breakable parts.

Set up several checkpoints along the way on the assembly line. Without good component parts, the final product is compromised. You must test the final assembled product to make sure that it performs to all standards. Your production system itself is the primary component of having a great quality control system in place.

When a product fails performance tests, it holds up shipments, increases inventory, and puts a severe financial strain on the manufacturer, as well as others along the way. When the quality of a product is poor, the waste and spoilage on the production line are greater than they ought to be.

For various types of products, mandatory government approvals must be complied with. Manufacturers must meet a variety of approvals and regulations to produce a product, as well as to be able to stay in business.

When a product is of poor quality, all departments and companies that carry the product suffer. For example, if you purchase a poorly made shirt, your opinion of the brand, and the shop that sold it, goes down.

If quality is so compromised that you have to recall the product, you end up paying big bucks. If you want to salvage any goodwill, you give customers a full refund, and with the costs of delivering that refund and replacing the damaged merchandise, you're considerably out of pocket. That's not how you want consumers to remember you.

Part IV
Preparing to Enter the Market

'It's our latest invention–saves us getting
in and out of the cab all the time.'

In this part . . .

It doesn't matter how great your invention is if you can't bring it to market effectively. A successful inventor either needs to become an entrepreneur, or find the right people to take his product to market for him.

Becoming an entrepreneur requires a whole new set of skills. These chapters help you understand the risks and rewards of starting a business to produce or sell your idea. They guide you on creating your business plan, finding funds, choosing associates you can trust, and getting ready to launch your product.

Chapter 14

Developing a Business Plan

. . . .

. . . .

*F*ailure to plan is planning to fail! A business plan is the first step you take towards turning your invention into a marketable reality. Like any other well-thought-out plan, a business plan goes a long way towards helping you succeed in your business. It encourages you to consider the important factors that affect your route to market, and gives you a map to follow.

Many potential investors understand a business proposal much better than they understand the technology of an invention. Business people usually look at the profit and loss possibilities differently to the way you look at them. They want to see your plan of action.

This chapter takes you through each section of a business plan and helps you to think about what type of information to include.

Why You Need a Business Plan

You have your great invention. You're taking steps to protect your intellectual property (Chapter 3 explains how to do this). Now you're thinking about how to make money.

Your first step is to write a *business plan* – a step-by-step, section-by-section account of your invention: how you plan to produce, market, and sell it; how you propose to raise the money you need; how you intend to make a profit (as well as how much and how soon); and who the key people involved are.

Sitting down to write a business plan may sound like a boring and unnecessary task, but you need to spend time and effort thinking about your strategy for success. If you don't have a business plan, you don't know how to get your business rolling or your product off the ground. Furthermore, without a business plan, you won't know what to do if your idea succeeds, and you can't take full advantage of a success you're not ready for.

If you plan to license your invention to another manufacturer (skip to Chapter 21 for more on licensing), working up a business plan helps you to consider all aspects of the deal and gives you a heads up on what to expect.

A business plan helps you set out your goals and objectives. And just saying 'I want to make lots of money and be rich' doesn't give you enough information. Until you determine how much money you want to make and formulate a plan to get some of those riches, you don't stand a chance of achieving anything. You face the tough job of figuring out your goals and objectives and how to accomplish them, but this book is here to help.

Setting goals and objectives is one of the most important steps you have to take to sell your invention. So you need to translate those goals into concrete action with a business plan. All the pieces of your business plan help you define your goals very clearly so that you can work out how to achieve them.

Working Up a Business Plan

Think of your business plan as a road map that shows you the way to your goal. Your map provides a step-by-step plan to build your invention into a profitable business, so you have a written guide to help you reach your goal.

Creating a business plan takes you through a number of steps in which you analyse each and every aspect of your idea, from creation to profit.

What your plan can do for you

A successful business plan demonstrates that you know your product thoroughly and that you've studied the industry and competition. Investors and clients consider a good business plan a primary indicator of future success with a well-organised company.

Businesses create many, if not most, business plans for the primary purpose of attracting funding. But you can also use business plans to attract business partners, advisers, and even potential customers. But first and foremost, a business plan keeps you on track.

Be clear about what you want the plan to accomplish and who you aim your plan at, and tailor certain sections to meet your goals or your audience's needs:

✔ A business plan aimed at raising funds lays out the amount of funding you require, what you offer in return for those funds, and the potential payback for the investors. A plan written primarily to get funding needs an emphasis on *financial projections* (projections of the money you need and outlay, considering the costs of goods sold and profit margins over time); *break-even analysis* (how many units you must sell before you cover the cost of your expenses); and *capital structure* (the type of business you have, how you fund it, and how you'll distribute the profits and losses).

Your audience probably includes private investors, venture capitalists, and other financial types who see hundreds of business plans. For them, your plan needs to show solid, realistic numbers and attractive potential.

✔ When you write a section of your business plan to attract experts in the field to become involved in the business, you focus on university affiliations, government regulatory approvals or applications, patent positions, and ongoing research and product development. You must write your plan very well in order to attract the kind of people you need for technical and professional advice.

✔ You may target a part of your business plan towards attracting a partner to manufacture your product or a marketing group to assist you in bringing your invention to market. The plan helps you attract those who can work with you along the way. Sometimes, these people work for you on a percentage basis. The plan lays out a path for their investment of time, resources, and money.

Who writes the plan?

As far as possible, put the business plan together yourself. You're the one with the idea, invention, product, or service, and you know more than anyone else about it. You know what you're selling, who your potential customers are, and why they buy.

Every buyer, bank manager, investor, manufacturer, and department store representative asks you questions that a well-thought-out business plan answers. The more detailed your plan, the better you can use it to respond to questions about your invention and business.

Inventors and entrepreneurs sometimes make the mistake of hiring a large consulting firm to write their business plan for them. When the inventor takes the business plan to potential investors and the investors start asking detailed questions, the inventor often can't answer questions regarding their own company and its projects, competition, goals, and plans.

If, for some reason, someone else does write your business plan, make sure that you're familiar with every aspect of it so that you can answer any questions that come up.

Breaking Down Your Business Plan

The following sections offer detailed information about what goes into a business plan. Figure 14-1 lists the sections of an effective business plan. Your plan may or may not include all these sections, but this template can help you get started.

Business Plan

Cover Page and Purpose of the Plan
Outline
Executive Summary
Market Analysis
      Industry Overview
      Target Market
      Market Test Results
      Industry Sales Cycles
      Competition
      Government Rules, Regulations, and Restrictions
The Company
      The Product (or Service)
         Detailed Description
         Product Life Cycle
         Research and Development
         Intellectual Property
      Marketing and Sales
         Marketing Strategy
         Distribution & Sales Force Creation
      Operations
         Manufacturing
         Human Resources / Personnel
         Physical Locations
Management and Corporate Structure
      Organisational Structure
      Key Management and Directors
      Legal Structure
Capitalisation
Ownership and Existing Capital
      Current Needs and Use of Funds
      Long-Term Needs and Use of Funds
Payback and Exit Strategies
      Financial Information
         Summary Financial History
         Summary Three to Five Year Projections
         Detailed Financial Analysis
Appendixes

Figure 14-1:
The parts of an effective business plan.

Introducing your business plan

The *Front page* of your business plan gives the name of your company and contact information. Often, the Front page also indicates the *Purpose of the Plan*.

Try to decide what the purpose of your business plan is before you write it. Occasionally, you can write the body of the business plan and emphasise a certain aspect afterwards, but keeping the goal in mind as you write certainly makes your plan more focused.

Sketching it out

If you send your business plan to a potential investor, they may already have 20 other business plans sitting on their desk. You want yours to be the one that they read. The next section talks about hooking the investor with the Executive Summary, but you get your first chance to spark interest in your proposal with the Outline.

Keep the Outline simple, and no longer than one or two pages; basically, it provides a table of contents. It tells readers what to expect and shows them that you give appropriate consideration to all the relevant areas.

Writing an Executive Summary

The Executive Summary is exactly what the words imply – a few pages that succinctly sum up your entire business plan. You state what your invention is and how you plan to bring it to market.

The Executive Summary is the most important part of your business plan. Potential investors read the Executive Summary first, so it has to attract every reader in its very first paragraph. Your Executive Summary must make readers immediately want what you offer, and want to invest in your business or get involved.

Think of the Executive Summary as the film trailer for your business plan. You want it to highlight the best parts of the plan in order to entice the reader into wanting to see the complete film. If the Executive Summary trailer is dull and boring, the investor may just get bored and walk out of the cinema before the film starts!

Along with your concise but captivating introductory paragraph, the Executive Summary sums up each of the major sections in your business plan. Imagine that your reader has only 15 minutes to glance at your business plan. Pick the highlights or most important points from each of the other sections and put them in the summary. For example, the synopsis for the Product (or Service) section answers the primary questions about why you developed this product and why its features are superior to the competition. The synopsis for the Capitalisation section explains the current ownership structure, the current funding needs, and the potential payback.

Revealing your research

The Market Analysis section of the business plan reviews the potential market for your product and your product's potential position in that market.

Research is crucial before you write your Market Analysis. If your product is a board game, for example, you need to assess the toy industry in total and figure out what portion of the spend in that industry goes to board games. You have to find out which other board games sell, both new board games and those games that have been around and successful for years. You need to look at how businesses market board games and design a marketing plan for your game.

You must know everything there is to know, not only about your product, but also about the industry in which your product competes. Work up a profile of your prospective purchaser and know how to make your product appeal to their needs. Making wild statements such as 'Everyone is going to buy my board game' is simply not professional. That scenario isn't going to happen, and such a statement probably sees your business plan heading for the investor's bin.

Refer to Chapter 18 to find out much more about researching the market for your invention.

Structure the results of your research in a format that can be easily understood such as spreadsheets, pie charts, and graphs. The best Market Analysis sections convince the reader that the market is sufficiently large for the prospect to have a chance of entering it successfully.

If the market for your product isn't especially large, you need to show that your product has a good chance of capturing a high percentage of the market that does exist.

Looking at the industry's big picture

The Industry Overview section is crucial to your market analysis because it shows the total market for your product and your product's potential within that market. The more information and data you show here, the more credible your overall plan appears to the reader.

Every product has an industry and any number of sources for data about that industry exist. Some of these sources include the government, the Internet, trade magazines, trade associations, and trade shows, along with annual reports from companies in the product line, and, of course, the people involved in the industry.

Your Market Analysis (see Chapter 18) shows the data you collected through your research of the industry. For example, if your product is an item for the skiing industry, you must show sporting goods industry sales as a whole because each pound spent on any sport other than skiing is a pound not spent on your product. Then you must show total pounds spent on items for the skiing industry as a subset of the total sporting goods industry. This comparison shows the potential size of the industry and the size of your specific market. Remember that you have to define your market by the product category for that product. Say you developed a new ski pole; you need to find out how many ski poles retailers sold in the previous market season.

Targeting the right people

As part of the Market Analysis, you need to indicate your *target market*, which means the likely end user of your product. In the Target Market section, you use the data from your Industry Overview to show who your actual consumer is likely to be. Continuing the ski pole example, you show that your target market consists of skiers in various regions of the world. Because nearly every skier uses ski poles (you're not interested in those daring but slightly crazy skiers who don't), you need to show the total number of skiers in the world by country. You also indicate how often skiers purchase ski poles – for instance, with every set of skis or not.

The more data you show about your target market, the more likely it becomes that your audience can see the potential of your product selling to that market. Again, you can gather much of this data from specific trade magazines and trade associations. The trade magazines for the skiing industry may show the number of skiers; the articles about ski pole manufacturers may show number of customers; and the trade associations may have data on the number of skiers who visit ski resorts annually. In addition, many advertising articles and magazines have marketing and advertising data that may be helpful to you. The number of subscribers to skiing magazines gives you a number of potentially serious skiers (though not the whole market). Magazines may also indicate advertising budgets, and if your research turns up the ad budget for a ski pole maker, you may discover that a certain percentage of total sales goes toward the ad budget.

Define your target market in as much detail as possible for the reader. You explain how you plan to capture part of that market in the Marketing Strategy section, later in the business plan.

Presenting the research results

After you define your target market, you need to do *market research* (see Chapter 18) in order to complete the Market Test Results section of your business plan. You conduct *primary market research* by talking one-to-one with potential customers. You can use a couple of different primary market testing methods to determine whether customers may buy your product:

- **Questionnaire:** You can survey potential customers with a well-designed questionnaire that indicates whether or not the customer would actually purchase your product.

- **Focus groups:** You can assemble focus groups specific to your target market and ask the group members for their opinions about your product and its features.

Secondary market research is existing data on a product or industry. You can look at market tests that other people published and use the data they collected on similar products. For example, if your invention is a new tent peg, secondary information is data on the number of campers, campsites, number of tents sold, cost of similar existing products, and the amount consumers spend on their camping equipment. You can gather data about a product that has many of the features of your product or about a product that you believe that you may compete with in the future.

Knowing your industry's cycle

Every industry and the products within that industry have marketing and sales cycles. The Industry Sales Cycles section needs to show that you're ready and able to fit in with the sales cycles so that your product has a fighting chance in the market.

In the European skiing industry, for example, retailers of skiing equipment order products for the next season from January to March. Providers deliver ordered products between September and late November, and the retailers sell those products from November to March. So if you intend to sell to the skiing industry, you must manufacture your product to meet the *lead times* (the time allowed to get the product through its distribution channels to the point of sale) for marketing and sales in January to March and then produce to deliver by November.

Clearly, you need to know and contract for relevant lead times to create a successful product and the reader of your business plan must be able to see that you take sales cycles into consideration.

Acknowledging the competition

To write a credible Market Analysis section, specify your competition and quantify that competition as best you can. Competition exists whether you acknowledge it or not. You're always competing for the consumer's cash, whether in a head-to-head battle with a similar product or in a more general lobby for consumers' *disposable income* (what's left after necessities such as the mortgage, bills, food, and so on).

The Competition section of your business plan needs to show why consumers should spend their hard-earned money on your product over all the other competing products. Your market research examines the target market, number of customers, and how often those customers consume a similar product. Outline each competitor, explain how the features and pricing of your product are better than that competition's product, and give the consumer good reason to buy your product. Many business plans contain a *competitive analysis chart*, which shows a table of every competitor and the features, pricing, service, and so on of its product.

Be thorough! You lose credibility if the reader of your business plan knows of a competitor that you don't mention. She's simply going to assume that you made other mistakes too and become disinclined to support you.

Abiding by the law

In writing the Rules, Regulations, and Restrictions section of the business plan, you demonstrate to your readers that you researched the appropriate regulations and complied with them, as necessary.

You must comply with the regulatory requirements for your product's industry in order to legally market or sell, or perhaps even advertise, your product. Health and safety provisions now affect almost every sphere of activity. For example, if you invent a new type of medical device it almost certainly needs official approval before you can sell it for use.

Most industries involve some form of regulatory compliance. Often, industry trade associations can help you determine what you need to do to meet any such requirements. Your manufacturer may also be able to help you figure out what you need to do. Check out Appendix B for a list of sources that may be able to help.

Introducing yourself

The Company section of your business plan sets forth your company's plans and describes your operating methods. You need to show that you understand what your business must do to succeed. You must demonstrate that you have the ability to locate customers and sell your product at a sufficient profit to stay solvent, at the very least. Of course, you fervently hope that you can make millions, so certainly show in detail that such an outcome is possible. Briefly describe your own short- and long-term objectives, such as reaching a turnover of £50,000 in year one, increasing to £500,000 in year five.

Unveiling your brilliant idea

The Product (or Service) section gives a complete account of how you discovered and developed your invention, and includes a very detailed description of the product itself – without disclosing any proprietary secrets. (Hopefully, you already applied for a patent, design registration, and/or a trademark to protect what you developed. See Parts I and II for more information on such intellectual property.)

Your product section needs to thoroughly educate your audience about your product. In doing so, you

- Discuss any technical advances that your product features
- Explain how and why your product is better than any similar product(s) on the market
- Prove that the product is worthwhile and predict what future it has
- Describe the history of your product's development
- State the problem that your product solves
- Identify the scope of applications for your product

Good visuals that show the product design and the project in action make up a key component of your Product Section. Include diagrams and photographs as appropriate, and if possible, refer readers to a Web site where multimedia features show the product in use.

Include information such as rough costs, packaging, price, size, and colour, and all technical specifications. Check similar products' descriptions and make sure that you make your product description complete. Go to shops that may sell your product and check out where those shops shelve similar products. See how much those products sell for and how manufacturers package them – the colours, size, and shape. Look at who does the manufacturing, as well as the distributing, of these products.

Answering 'What in the world is that?'

You may find the Detailed Description section of the business plan the easiest to write. Why? Because here you just describe your product, and you can give that description better than anyone.

To write a detailed description of your product, you simply need to provide answers to questions that readers may have:

- What exactly is your product?
- What does your product do?
- How does your product do it?
- What are the functional specifications?

✔ Have you tested the product to make sure that it works?

✔ If a prototype exists, who constructed and tested it?

✔ What are the product's size dimensions?

✔ How much does the product weigh?

If you kept a journal while you developed the product, use a few of the interesting points in the development process to describe your product's function. If you have a technical product, describe the technology involved, why that technology works, and how you can maintain your developmental lead.

Keep in mind that the person reading your business plan probably isn't an engineer, so don't get so technical that you bore or confuse them. In fact, most people don't care how the product does what it does. They care that the product does what you say it can do. For example, most people don't care how a mobile phone works, they just care that the calls go through!

You may want to include a chart describing each function of your product compared with the functions of similar products, much like those you find in consumer evaluation magazines. List all the features available, and then compare your product and your competitor's in each feature category. Also have a section for price comparisons.

Describing the life and death of your product

Every product has a *life cycle*, which we typically define as how long the product functions under normal use and conditions. The life cycle dictates how often a consumer has to repurchase a product. In the Product Life Cycle section you need to describe your product's life cycle, which, like a chain, is only as good as its weakest part.

The *planned obsolescence* theory holds that some manufacturers build a shortened life into their product to hasten the day the consumer must replace it. The opposite holds true too. The first Honda Civics ran virtually for ever if you changed the oil, making their life cycle longer than every other car of that period.

Consumers now know the life cycle of most products and industries well. Build to meet or exceed this life cycle and utilise parts that help you meet it. You don't need an expensive part that lasts 15 years for a product that you design to last for 5 years.

Consider life cycle as you develop and enhance your product. Customers always look for something newer, cheaper, and with better features at the end of the original product's life cycle.

Labelling failure a success

Researchers at 3M were trying to develop a type of adhesive paper that would stay stuck firmly to all surfaces. They thought they'd failed when the notes they developed didn't stay permanently stuck. The research department threw them out. Fortunately, the accounting department found them and started using them. They wrote notes and stuck them to papers, easily detaching those notes later. And thus, Post-it® notes were born.

Explaining how you'll continue to grow

Initially, research and development generates your new product. Research discovers a solution to a problem that leads to the development of a new product.

In the Research and Development section explain how you used research and development to create your new product and how you intend to continue to enhance your current product, and to find new products. You want to demonstrate that you have ideas for future growth and for the enhancement or replacement of your product.

Claiming what's yours

Investors and lawyers study the Intellectual Property section the most closely. Although they may not understand the process or technology you used to create your invention, they certainly understand if you've applied for, or acquired, a UK patent, registered design, or registered trade mark, and even more so if you also applied for, or acquired, equivalent rights in other countries.

Protection of your intellectual property increases the value of your product. A granted patent secures market protection for up to 20 years, provided that you pay the annual renewal fees. A granted patent also shows that you have a unique product that the law protects. Refer to Chapter 3 for more information about intellectual property.

Planning to reach, and keep, the customer

Your business plan audience knows that no company becomes successful overnight, but they want reassurance that you have a realistic plan for making money. The Marketing and Sales section of your business plan demonstrates that you know the potential customer profile (including age, gender, income, and decision criteria – why would certain types of customers buy your product over competing products?) and how to reach those customers, as well as your competitors, and the typical growth patterns of companies in your industry. It must also show that you're thinking beyond just making a quick buck.

The Japanese company Mitsubishi takes a very long-term view; its planning documents cover the next 1,000 years. You must also take the long-term view. Tell readers where the company is going to be in the next one, five, and ten years.

The marketing language in this section reminds the reader of the pertinent data you present earlier in the Market Analysis section and shows how you plan to make use of that data to best sell your product.

Your product needs the Marketing and Sales plan so that you can actually sell it, and therefore what you outline here must be convincing to you and your readers. Turn to Chapter 19 for more information on marketing.

Crunching the numbers to find the best bargain

A marketing strategy is choosing the best, most economically feasible method to sell your product, including how you'll advertise and who'll sell and buy your product.

In the Marketing Strategy section of your plan, first remind readers of all the valuable data and observations that you provide in the Market Analysis section. Then show readers that you can make tough choices, based on your analysis of this data. For example, say your data suggests that more elderly people buy your product than young people. Part of your Marketing Strategy section explains that you plan to advertise in the crossword sections of newspapers rather than on children's television because of the readership age.

The marketing plan shows that you know your customer, your potential sales, and at what price point your consumers agree to purchase your product. You must show your plan to enter the market, the trade shows where you'll exhibit, and your knowledge of the global market for your product. You must show how you plan to achieve sales and by what methods.

Using others (in a good way)

In the Distribution and Salesforce Creation section, walk the reader through how you intend to actually sell your product. Address issues such as:

- **Distribution:** Every industry has specific and well-defined distribution channels. Working within these channels usually gets your product to market faster and more easily than building your own distribution network, which is a time-consuming and costly undertaking. Tell readers which channel works best for your industry and your product, keeping in mind that companies often don't accept a product that falls outside conventional distribution channels.

- **Salesforce:** Tell readers whether an established salesforce that sells similar products represents your product or whether you're building your own sales department. Then convince your audience that you made the right choice for your product.

If you plan to license your product to an existing company within the industry (see Chapter 21 for more on licensing), tell readers why you made that choice and explain the licensing company's strategy for your product.

Explain clearly how you plan to use suppliers, distributors, and wholesalers both as channels of distribution and as customers.

Explaining how your business runs

The Operations section outlines the day-to-day operations of your business and shows the reader how these operations make the company profitable. You need to discuss the tasks of each department and show how the departments relate to each other and to the outside world.

The Operations section of your plan also details how many people you need to accomplish each task and whether you plan to hire people to work in-house or plan to outsource a number of operations – a customer service telephone line, for example.

Use this section to point out any competitive advantages your operations give you. You can find a competitive advantage almost anywhere inside your company – your employee base, your manufacturing process, your management team, your organisational structure, or even your reputation. You may have seen Hewlett Packard's (HP) advertising campaign that shows the HP logo next to the word 'invent'. HP is trying to suggest that it has a competitive advantage in the fast-changing technology world because HP employees are inventors, not followers. If this mindset really exists in your company, it can certainly be a competitive advantage.

Putting your product together

The Operations section of your business plan can address all types of processes that you carry out within the company, but it most commonly discusses manufacturing.

Don't think you can skip the Manufacturing section if you offer a service and not a product. Rather than describing actual nuts and bolts, you describe the figurative nuts and bolts you need to manufacture your service – the personnel, facilities, and so on that you need to make your service operational.

Even if your invention just improves an existing product and you intend to license that improvement to a third party, you need to explain how your invention adds value by enhancing the existing product.

Introducing the people who make it all happen

Human Resources/Personnel is an important subsection of Operations. You explain your company's organisational structure in Management and Corporate Structure (see the 'Presenting the corporate bigwigs' section, later in this chapter), but this section addresses personnel from a raw operations standpoint.

Given what you describe about manufacturing, research and development, and sales and distribution, how many people do you need and how soon?

You may have a tough time writing the Human Resources/Personnel section. You need to demonstrate that you researched the labour market from which you plan to draw employees, to determine competitive salary ranges for the qualifications you need, the benefits your employees expect, and how much those benefits cost. You need to explain whether you plan to have employment contracts with your product engineers. Set forth your thoughts on employee motivation techniques, whether you plan to offer commission or bonuses, how often you intend to carry out employee reviews, and where and how you can find employees. Even if you're the only employee, show the reader that you're thinking ahead to a time when your successful business expands.

Bragging about your digs

The Physical Locations section can be short (depending on your business idea), but don't leave it out. The smallest company has a headquarters (even if it's only your garage), and you need to tell readers where that office is, what relevant equipment it already has (photocopier, fax machine, phone line, broadband, and so on), and what it needs. Show your readers that you've thought ahead. You can include a paragraph that starts 'When the company reaches £X million in sales, it'll need to expand to a new location because . . . '.

Some businesses need larger versions of the Physical Locations section than others. If location is key to your business's success, you need to present an in-depth discussion of how you chose or intend to choose locations for business.

Presenting the corporate bigwigs

Management and Corporate Structure are so crucial to a business's success that many investors refuse to invest in a company that doesn't have an experienced management team. Business history shows that a good management team is just as important as having a good product. When making judgements on what to invest in, many investors choose the person or team behind the product. You can work with a great entrepreneur who has a good product much more easily than a mediocre entrepreneur with a great product. That mediocre entrepreneur may cause many problems in the long run.

You need to show that your organisation and management team have the capabilities to carry out the purpose and achieve the goals of your company.

Displaying your team

Your Organisational Structure section shows the chain of command and responsibility for specific functions within the company. Often, an organisational chart depicts the company's structure and all the functional responsibilities within that company.

The structure of a company depends on the type of business operations involved and the industry in which it operates. A manufacturing company, for example, needs manufacturing and product development departments, along with the standard accounting, finance, human resources, marketing, and sales departments. The economy also plays a role in business operations. Internal structures become smaller and tighter in difficult economic times, and expand in periods of economic growth.

Even if you don't include an organisational chart, make sure that you provide a clear description of how information flows within your company. Direct, formal communications as well as more informal company newsletters or weekly staff meetings build morale and inform employees how they can do their jobs better.

Don't forget to also stress the importance of external communications to customers and suppliers.

Identifying the players and their pieces

In the Key Management and Directors section of the business plan you need to discuss the key players you already have on your team, how you rely on them and what you expect of them, and the spots you want to fill. The *key players* in any organisation are those experienced people who can get specific tasks done, for example:

- ✔ The inventor or entrepreneur – probably you! – is a key player, especially at the start. Other key personnel may come on board as the company grows.
- ✔ A well-known scientist may be a key player in an organisation selling a medical device.
- ✔ The person attracting capital to the venture is often a key player.

Key players also usually have a network they tap into for advice and help – they know other key players in the industry.

Placing a key player within the management of your company is a smart move – they add experience and substance to the business. Using key players strategically is a good idea. For example, if you intend to apply for research and development grant funding, you may consider finding a key player who was awarded a similar grant and is familiar with the process.

Make sure that you include the biographies of your key players or team members in synopsis form in this section, or as part of the appendixes. Also, disclose any family relationships between key players. If you and your spouse are the two key players, investors can infer that you're skilled in managing at least one relationship. Also mention here any employment contracts that you secured with your key players.

Explaining the structure of your company

In the Key Management and Directors section of your business plan you need to indicate the current arrangement or 'legal structure' of your business. Are you working as a sole trader, or in partnership with someone else, or have you already set up a limited liability company to support the development of your invention? (See Chapter 16 for information on the various types of business structure.) If a limited liability company is already in place, provide information on the share structure, when the business was incorporated (including the registered office and company registration number), and identify the director/s and company secretary. You may like to explain why you chose the name of your company and if you registered it as a trade mark (see Chapter 8 for more information on registering your trade mark).

Get everything settled before you write your business plan, though you may want input from your adviser on the language you use in the Legal Structure section of your plan and how to explain why your decision is the right choice for your company.

Considering the money

The Capitalisation section of the business plan describes the funding behind your idea. The detail that this section requires varies greatly depending on the purpose of your plan. For example, if you're sending the plan to a venture capitalist (see Chapter 15) or other potential investor, you provide detailed specifics, such as ownership percentages. If you're sending the plan to a potential partner, you can make the Capitalisation section a little more general, letting readers know that you're the major shareholder and mentioning other relevant shareholders.

Identifying who puts money in and who takes money out

In the Ownership and Existing Capital section, especially when you present to potential future investors, include a table showing all the owners, with the amount of money each invested and what percentage of the total ownership each possesses. You may want to show your fundraising time-plan – 'The two founding partners contributed £30,000 each in January and raised an additional £25,000 from family members in March', for example.

Provide a paragraph for each loan (including property) the company has outstanding. This section details to whom you owe the loan and the terms of that loan (particularly interest rate, payment, and collateral requirements).

Outlining what you need now and how you plan to use it

The Current Needs and Use of Funds section lays out all the company's funding needs for the next six months. You present the budget for each department and state how much money you need to meet your obligations for payroll, taxes, rent and utilities, production costs, marketing and sales expenses, and everything else the business has to spend money on.

Consider using an accountant to help you put the cash flow projections together, and make sure that the financial director of your company reviews it.

This section tells readers how you plan to use funds. Draw up a table showing investors how you plan to utilise the capital they invest and what results the company hopes to achieve in production and sales. You don't have to explain why you need the amount you do for a specific area but the amounts should make sense to the reader if you do a good job in the other sections. For example, if you manufacture an electrical component that requires very expensive raw materials, readers won't be surprised to see that the price in the raw materials category is high.

Looking for funds farther down the road

In the Long-Term Needs and Use of Funds section, you put forth the company's long-term capital requirements. You provide three- to five-year projections in the Financial Information section but you use the Long-'Term Needs section to address where you expect your funds to come from. You typically include a sources-and-uses table that shows where you expect to get your money and how you expect to use it.

All the figures in this section must match up with the figures you provide in the long-term projections in your Financial Information section.

Getting even and getting out

The Payback and Exit Strategies section details how and when investors get the return they seek on their investment. You summarise the amount of capital you seek and the amount others already invested. You may also remind readers of your long-term projections – 'If things go as planned, we'll reach the break-even point by the end of year two and have cash flow of £X by the end of year five', for example. Then state the projected payback to investors.

Investors want to know when they can expect you to return their capital to them, and you need to present an appealing *exit strategy*, which is another way of saying when you expect to be able to return their investment and at what level of profit. They don't want to be married to you and your product forever, and you don't want their input for ever – you *both* want an exit plan.

The details of your proposed exit strategy often determine the structure of the original investments. An investor may loan the company money with a specific interest rate and payback date. Alternatively, the investor may loan money with an option for the investor to convert the loan to shares in the company instead of repaying the investor.

Seek independent advice from a financial and legal consultant knowledgeable in the area of Payback and Exit Strategies.

A smart reader may check the Financial Information section to judge whether you've a reasonable chance of turning your idea into a reality. You want to make this section optimistic (you believe in your idea, don't you?) but realistic. The more work you put into the previous sections of the business plan, the easier writing this section becomes.

Telling your financial story

The company must have good accounting records and financial history. Typically, the Summary Financial History section includes past and current financial statements, tax returns, and income analysis for each year the company has existed. If you're a brand-new company and don't have past financial records, just say so.

Companies fortunate enough to show a number of quarters or years of increased earnings happily show their financial history, but even those companies with less impressive results must show their past history (though you can provide discussion of any obvious setbacks the figures show).

Keep in mind that readers compare your financial history with your current funding request and projections, so make sure that the numbers are reasonable.

If you have intellectual property in patents, designs, or registered trademarks, you can assign a value to it here. The Summary Financial History also lists any professional firms involved with your accounting and financial management.

Predicting your future

The Summary Three- to Five-Year Projections section has become a requirement of the fundraising process. The projection of expenses and revenues often gives potential investors a good idea of what you believe potential revenues to be. A five-year business plan also helps you plan where you want your company and product to go in the near future.

In some cases, you may want to include multiple scenarios here. For example, you may be trying to break into both domestic and foreign markets. In such a case, you may want to show what may happen if you enter just the domestic market. Or consider the case of a product that's very dependent on outside factors – you may want to show separate scenarios for favourable and unfavourable conditions. If, for example, you invent a new raincoat fabric, you can forecast what may happen if your market area receives record rainfalls and what happens in an unusually dry winter. What you're really trying to do with all the sections is to demonstrate that you're presenting a well-thought-out business plan, covering all contingencies.

Making sure that the numbers match

The Detailed Financial Analysis section reviews the projections, any historical and current financial data, and the sources and uses of proceeds. The information in this section compares what you say in each preceding section of your business plan with the others for consistency.

The Detailed Financial Analysis section also provides you with an opportunity to explain any costs or projections that fall outside the norm. You can explain your assumptions about the raw material cost of each of your products or why you project that next year's travel and entertainment expenses may actually go down.

Tacking stuff on at the end

Reserve the Appendixes section for data or materials pertinent to the success of your idea that don't fit into an easily readable format. Very detailed documents that have too many pages to put in the body of the business plan often end up here. You generally locate the following types of documents in business plan appendixes:

- ✔ **Advertisements:** If you've already produced advertisements for your products or services in print form, including a sample is always a good idea.

- ✔ **Contracts in hand:** Include any contract, purchase order, or letter of potential use here. Contracts and purchase orders provide valuable support when backing up projections. If you want to keep your clients' names confidential, you can block out their names or assign them generic names (client A, client B, and so on).

- ✔ **Employment and contracts:** If you recruit valuable personnel include your employment contracts with them that show how long you expect them to work with the company and under what rules and regulations. Also include clauses signed by key personnel that prohibit those personnel from competing with you and from taking product concepts to another company (see Chapter 11).

- ✔ **Insurance policies:** Most investors look for general, key person, and product liability policies to ensure that you're protecting the company's assets.

- ✔ **Lease agreements:** Include copies of any leases for space or major equipment.

- ✔ **Owner agreements:** Include any agreements among the owners of your company involving equity in the business, the business property, or any other business-related issue.

✓ **Patents, registered designs, and registered trade marks:** Consider including copies of your intellectual property documents to allow readers see what you've accomplished and to save them the time of getting these documents from the intellectual property office.

✓ **Pictures or recent presentations:** Photographs or a good PowerPoint presentation really help you get your point across. For example, pictures can show the management team in action, the product's production process, and the company's physical location.

✓ **Research statistics:** Include all the results of your research studies, including both primary and secondary research on your product's potential market. This appendix may be very long, but it also may be at the heart of your technology or scientific invention.

✓ **References:** You can include detailed references, including personal and business references with their contact numbers. Also include any testimonial letters about your product.

✓ **A short descriptive summary of the management team and board of directors:** You can place this information here or choose to include biographies in paragraph form in the Key Management and Directors section (see 'Identifying the players and their pieces', earlier in this chapter).

What you include in this section is really up to you; think about who you want to read the plan in order to decide what you can't leave out.

Chapter 15

Finding Funding

*Y*ou can sum up this chapter's theme with the simple saying 'It takes money to make money'. You may have the greatest idea in the world, but without some sort of financing, reaping the rewards of your idea is next to impossible. Although poor management most commonly causes businesses to fail, inadequate or ill-timed financing runs a close second. Whether you're starting a business around your new invention or expanding your product line, sufficient ready capital (in other words, having enough money readily available) is a must.

The success of your business rests on having enough money to keep that business running smoothly. Your cash flow affects all aspects of the company, from operations to your ability to deliver a final product or service. But having sufficient financing isn't enough; you have to be knowledgeable about money too. Securing the wrong type of financing, miscalculating the amount of money you require, or underestimating the cost of borrowing money can all hurt your business.

This chapter describes the various ways that you can obtain the required funding for your invention or your business, alerting you to the risks associated with each option for you and the investor.

For many inventors, raising capital isn't a favourite task; they want to be inventing new products or services. If you identify with that philosophy, consider hiring or partnering with someone who's good at asking investors for money, to keep on top of your funding needs. A big difference exists between an inventor and an entrepreneur. Investors and companies tend not to invest in inventors; however, they do invest in entrepreneurs.

Determining How Much You Need and for What

You certainly need money to launch your invention into the marketplace. But before you contact investors or apply for a loan to fund your business expansion, ask yourself whether you really need more money or just a different strategy.

Considering cash flow

All your funding needs relate directly to your company's cash flow, the most critical element of every company's day-to-day operations. *Cash flow* is the amount of money you pay to meet your expenses (acquiring raw materials to manufacture; supporting administration, marketing, and sales personnel; and so on) and the money you get from sales of your product or service. Insufficient or erratic cash flow can cause financial problems, and it can also result in poor company morale (employees worrying about whether you'll pay them tend not to give their all). Many funding needs come from a cash flow problem.

Understanding your basic funding needs

Before starting a search for funds, know how much you need, why you need it, when you need it, and what you need it to do for you. Before trying to find funding, answer the following questions:

- ✔ **Do you really need more money, or can you manage what you have more effectively?** Do you know what you really need to get started? Have you sat down with a pen and paper to figure it out, or are you just assuming that you automatically need more money from credit cards, investors, or bank loans? Take the time to plan a strategy and look at your options; whether you're going to sell your patent rights and license your invention or whether you're going to start your own business. Consider your personal strengths and weaknesses and talk with your partner at home about their dreams and vision as well. Sometimes you can cut corners in your spending habits without going into debt or automatically seeking partners and investors when you first get started.

- ✔ **Why do you need money?** Are you trying to introduce your product to market, expand your range with a new product, or do you simply want a financial cushion? When investors invest in a product, they want to know how you plan to spend their money. They want to make sure that you aren't going to buy a new car, build a new house, or take an exotic holiday. They need to know that you'll spend their money wisely.

✔ **How urgent is your need for funding?** Do you need money to meet immediate obligations and, therefore, have resigned yourself to paying higher financing costs, or can you wait and negotiate favourable lending terms?

✔ **How do you plan to use the money?** Are you trying to make a prototype, hire people, pay distribution costs, or expand your product line? Realise that your need for funds is most critical during your product's transitional phases. You may think that you need money in the start-up phase only, but find that you need funding in different stages as your product moves through its life cycle. Also realise that any lender requires that you request the money you borrow for very specific needs.

✔ **How does your need for financing coincide with your business plan?** From bank manager to business angel, they want to see your business plan. If you don't have one, write one! (Chapter 14 guides you through writing your plan.)

Looking at what you need money for

Before you go running off to talk to a loan provider or some other financial type, take a close look at your current financial situation and think specifically about what you need money for.

If you're looking to start a company to make and sell your invention, you need a detailed breakdown of all your expenses, from renting or buying space to hiring workers and getting your product to the customers. Perhaps a new assembly process can turn out your product more quickly and efficiently (and thus more cost effectively), but you don't have the funds to pay for the new technology right now. Maybe your business is growing, and you need help with the initial staff costs after adding more workers to handle the load.

Whatever you need or want money for, make sure that you've a detailed explanation of how you plan to use the money and how that plan ends up making more money for your investors. Even if you get the money from family or friends, you need to reassure them that you're not just upgrading the sound system in your car.

Sometimes taking a close look at your financial situation shows you that you don't really need more money at all. Delving into your income and expenses may reveal that you're spending money on a machine that you're really not using or a service that you can do yourself. You may find that you can remove from the budget an expense that made sense at one time but isn't necessary any longer.

Determining how much you need

Logically, how much money you need depends on what you need it for. If you're looking to start a new company you probably need more money than if you want to build a simple prototype.

Unfortunately, this section doesn't include a handy list that tells you that you need £X to start a company from scratch, £Y to expand into a new market, and £Z to advertise your invention. You have to find the numbers by researching into the going rate for the services or supplies you need.

Doing the explorations, market research, and comparisons you need to write a business plan gives you probably the best way to figure out how much money you need. Some investors and lenders require a business plan, and even if the financial types don't need to see that plan, working one up familiarises you with all aspects of your business. If you haven't written a business plan yet, do it now. Chapter 14 walks you through the process step by step.

Give yourself some margin of error (whatever you feel comfortable with) when you figure out how much cash you need – don't underestimate it!

Figuring Out the Types of Financing

Don't worry about the ins and outs of convertible debt, dividend protection rights, preferred equity, seed capital, put options, or other unintelligible topics. You just need enough financial knowledge in order to understand the language your financial advisers and backers speak. The big picture issues are pretty simple, and as the person in the driver's seat, you need to have a grasp of them.

This section defines and briefly discusses the various financial terms you're likely to run into, and describes how investors rate risk and return when deciding on whether to fund a company or project.

Defining the terms

The two main types of financing include debt financing and equity financing. *Debt financing* is like taking out a loan – you promise to repay the money within a certain time period and with a certain amount of interest. With *equity financing,* you basically sell shares in your company – you share your profits (and losses) with the investors. Debt financing poses the lowest risk for the investor (along with the lowest return potential), and equity financing poses the highest level of risk for the investor (but with an infinite return potential). Other forms of capital fall somewhere between these two forms. (For more on risk and return, see 'Rating risk and return', later in the chapter.)

The following list defines the types of funding options available for debt financing and equity financing:

- **Secured financing:** Like a loan guaranteed with collateral, an asset backs a secured debt. If you don't pay off the debt, the investor keeps the asset, so a secured debt is a low-risk investment with a low return rate to match.

- **Unsecured financing:** Riskier for the investor than secured debt, unsecured financing is basically a loan without collateral. For the higher risk involved, the investor expects – and gets! – a higher return rate.

- **Convertible financing:** This form of financing starts out as a regular loan-type arrangement, but if the company does well, the investor may choose to convert some or all the loan amount to shares in the company (that's where the *convertible* part of the name comes from). The risk factor here is sort of 50/50. If the company does well, the investor converts the debt to shares and potentially reaps large rewards; if the company does badly, the investor probably gets the investment amount back with just a small amount of interest.

- **Ordinary shares:** These are shares in a company that give the owner a vote in how the company runs and entitles the shareholder to receive dividends. Probably the riskiest investment type, common shares also have the most potential for a spectacular return.

- **Preference shares:** These are shares in a company that pay the investor a fixed dividend before the company pays dividends on ordinary shares. This investment has a fairly low risk rate – after all, preferred shareholders get their money before anyone else does. But the set dividend rate means that the investor doesn't benefit much if the company does really well.

Any form of debt is an obligation that the company must pay back in the future. You show what you owe in the long-term debt section of the company's balance sheet, and investors consider all debt a liability. Debt financing is a costly form of capital because you have to pay the interest, in addition to the capital you originally borrowed, and you often have to decrease your cash flow to meet your repayment deadlines. Too much debt can burden your company because paying off the debt uses cash that you can't use for other operations. A potential lender looks at the ratio of debt to equity as a barometer of a company's financial health. Too much debt sends up a red flag for a potential investor.

When looking for money, you must consider your company's *debt-to-equity ratio* (the comparison between the money you've borrowed and money you've invested in your business). The more money you personally invest in your business, the easier attracting additional financing becomes.

Rating risk and return

To get a good idea of how to look at your capital, funding, or financing (terms we use interchangeably in this book), think like the investor. If you had money and an inventor like yourself asked for some of it, what questions would you ask? These three questions drive every financing structure you encounter in raising money:

- ✔ When can I expect to get my money back?
- ✔ How much am I going to get back?
- ✔ What happens to my money if things go wrong?

The answers to these questions determine the level of risk for the investor and the conditions you place on yourself.

Table 15-1 answers investors' main questions for both debt and equity financing.

Table 15-1	Comparing Debt and Equity Financing Returns	
Investor's Question	*Debt*	*Equity*
When can I expect to get my money back?	A specific time in the future	Uncertain
How much am I going to get back?	A fixed amount over the original investment	Uncertain
What happens to my money if things go wrong?	You usually get some of your money back	You'll be lucky to get any money back

You're thinking that the equity column in Table 15-1 looks a little thin. An equity investment – buying shares – may or may not pay off. So why would you buy shares in the first place? Well, you probably already know the answer: because you hope for a high return. The *most* you can earn by giving someone a loan is the loan amount and whatever interest you agree to. But the same amount invested in shares in a company that takes off may theoretically yield an infinite amount of money (or at least more than you'd know what to do with).

Just as you can guarantee a loan with *collateral* (something of value that you forfeit if you don't pay the loan) or sell the same kind of shares to all your investors (*ordinary shares*), you can choose (or the investor chooses) from a range of financing options with different levels of risk for the investor and obligations for you (see the earlier section 'Defining the terms' for details on your various options). Figure 15-1 shows the range of options on a scale from low risk, low return to high risk, high return.

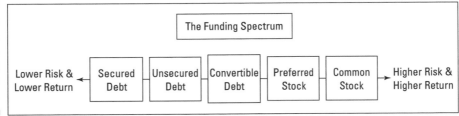

Figure 15-1: For investors, low risk means low gain.

 When starting a company (so you can market your product) you probably want to seek debt financing in the form of loans if your company has a high debt-to-equity ratio. However, if your company has more equity than debt, increase your ownership capital (equity investment) for additional funds. The more funding that you personally put into your product, the larger your percentage of ownership in the company. Remember – ideas are cheap; products are expensive.

 Make sure that you get proper legal advice before you attempt to sell shares in your venture.

Seeking Out Sources of Capital

You can go to any number of sources to finance your business – banks, commercial finance companies, individual investors, venture capitalists, and more. In addition, in recent years the Government has developed regional schemes to encourage the growth of small businesses, because these build the local economy and create jobs. This section covers the basics for each source.

 You're doing nothing wrong by asking people to invest in or lend money to you. Many people are looking to do exactly that! Venture capitalists exist to fund new businesses. They root for you to succeed so that they make money in their investment. So, yes, their interest is a little self-serving, but as long as it serves you too, you both come out winners. Even successful companies that go public and generate billions of pounds in revenues still have cash flow problems sometimes and seek to borrow additional funds.

Identifying the most important source – you!

If you're like most inventors, your primary and initial source of capital is your personal financial resources. You may consider using personal savings to finance the development of your invention.

Think very carefully before using a credit card or a home equity loan. Both involve risk and don't provide long-term solutions. If your company or product fails, you still owe the credit card company or mortgage company and may not have the cash to pay back what you owe. On the other hand, not putting your own money into your company tells potential investors that you aren't willing to take the same risk that you're asking them to take.

All your company's investors and lenders, even suppliers and customers, want to see that you, as the inventor, put your money where your mouth is. You may feel that the years you spent on your product should count as your investment. All inventors have this so-called *sweat equity*, and unfortunately for you, sweat equity investment doesn't carry a lot of weight with investors because they want to see you put your own cash, as well as time, into your invention.

If you want people to invest in you and your product, you have to show them that you're confident enough to put your own money at risk. If you're not willing to risk your money, why should they?

Be smart about investing in yourself and evaluating how much money you can risk. Keep in mind both your business's day-to-day start-up cash needs and your own personal needs. Borrow from your lowest cost sources first before charging everything to your credit cards and paying astronomical interest rates.

Keep track of all the money that you invest in the company (even £15 for an ink cartridge that you use to print business plans). You want your business to pay you back one day. After all, you put this money at risk, so you should get a return on it.

Getting a little help from your friends

Friends and family, your sports team mates, former colleagues and acquaintances, and fellow book club members – start looking at all of them for your start-up money. You find it easier to test your idea out on your closest friends and get their criticism than approaching professional investors. The same theory of personal investing applies here: If you can't convince your sister that your business is going to work, how can you ever convince somebody else?

Be very careful! You see your non-professional investors, like your friends and family, for the rest of your life. And you don't want them to hate you if you lose their hard-earned investment.

You may also receive more advice than you need from people who aren't shy about commenting on any or every aspect of your business. Some of their advice may be helpful – after all, your family and friends may have useful contacts – but ask yourself if you can deal with your retired Uncle Dave showing up at the office every week to tell you his latest idea about how to run the company.

Going the conventional route with bank loans

Most businesses borrow money at some point in their life cycle, and traditionally, small businesses use banks as their major source of funding. The bank's principal role is that of a short-term lender, offering loans for buildings, machinery, and equipment, and providing lines of credit for seasonal or unexpected expenses.

Banks are reluctant to offer long-term loans to provide the full cost of starting a small business, though you may find a bank offering an attractive lending option after your business is up and running and has a proven track record. At that point, you can probably get a loan without using personal assets to guarantee it. In fact, banks may compete to offer loans to you. Shopping around for the best rates and terms makes sense.

You can try to negotiate a holiday period at the beginning of the loan to start making repayments at a later date.

Different banks manage applications for conventional business loans in different ways. If you apply for a relatively small loan, the business manager at your local branch may authorise your application, in which case you need to prove the strength of your business face-to-face. For larger amounts (typically over £10,000) your application probably goes to a central business-lending unit for assessment. In this situation you probably won't have the opportunity to be there in person, so your business plan needs to stand up for itself.

Taking out a business or a personal loan means putting your personal assets at risk. Often, a bank lending to a start-up business requires collateral, such as your house or personal property, to guarantee the loan. If your company can't repay the loan, the lender comes to you to make good on it. You may have to forfeit the collateral you pledged or pay off the loan according to a schedule the lender sets up. So, before you apply for a loan, do a careful assessment of the risks involved. Can you handle the worst-case scenario?

Interest payments mean cash out of the door today that you never see again. Don't borrow so much that you can't make your current cash payments. When you get in over your head, the dream dies. If you can't seem to get sales of your product going or establishing a market share, sooner or later sales start to decline. The lack of sales added to increasing debt payments make businesses close their doors.

Make sure that the bank sees a well-run, well-managed business that's executing an up-to-date business plan.

If you don't understand a loan agreement then have an experienced professional read the fine print before you sign it. The documents are usually pretty standard, but you particularly want to know what kind of rights the bank has if you don't make your interest or principal payments on time.

For small- to medium-sized businesses operating in the UK, the Small Firms Loan Guarantee scheme (SFLG) is designed to provide access to funding where a conventional loan is not available. The SFLG scheme gives specified lenders a loan guarantee from the government to help qualifying businesses obtain funding of up to £250,000 with a low interest rate. For more information, visit the Business Link Web site (www.businesslink.gov.uk).

Calling on angels

Private individuals who invest in businesses are known as *business angels,* and who doesn't want an angel involved in a new enterprise? Angels are typically successful inventors, professionals, and entrepreneurs that have money to invest. In addition to financial support, these private investors may offer to share their knowledge, experience, and contacts, so that you get even more for their money.

Planting the seeds of invention

Seed capital or *seed finance* is early-stage, high-risk, and high-cost equity. Seed capital typically comes in the form of early stage funding from an angel, individual investors, and occasionally venture funds. The investor gets ownership in your company by receiving equity, and you don't have to repay the funding that comes into your company. Investors consider their investment *paid-in capital*, and that investment becomes an asset of your company.

You want to obtain seed capital investment rather than debt capital, if you can, because obtaining funding that you don't have to repay shows a greater commitment toward the company. If you can't find seed capital and debt capital is all that your company can obtain, then you obviously have to go with debt capital.

But remember, seed capital is so called for a reason. It gets the plant from the soil to the sunlight, but it's not the end of the funding story.

Private investors come from every industry and all walks of life. But they all have money and extra capital to invest. You can find them through your bank manager, your local Business Link, or possibly your accountant. You may find an angel or two through an inventors club, a local business people's networking group, or any other organisation with a financial aspect.

Don't be shy about letting people know that you're looking for investors. Most private start-up funding comes from people connected to, or referred to the business's owner. Money is crucial to business – everyone knows that. Ask anyone and everyone for money and contacts. Tell people about your product and your company and ask if they know anyone who may invest. (And always have a copy of your business plan available to pass along.)

Even if a potential investor says no, that non-investor may give you someone's name or offer to contact a person who turns out to be your biggest backer. So make sure that you follow up on any referrals.

Angel investors and venture capitalists seek to add value, in addition to capital, to the companies in which they invest so that the company grows and the investors achieve a greater return on their investment. To accomplish this big return, these investors often insist on being actively involved in the company. And almost all angels want, at a minimum, a seat on the board of directors.

Paying the bills on your time

As your company grows, you may find your cash flow situation tightening up – your customers pay you slowly and you have to pay your bills quickly. Don't worry: Quickly growing companies often have negative cash flow – it's a perfectly normal situation.

But if you can arrange to pay your creditors after you get the money from your customers, you can ease the cash flow squeeze. Industry competition often dictates when you have to pay your bills or when your customers pay for your products. If you sell to a giant distributing company, the company may tell you that it pays invoices 45 days after receiving them – and it's just too bad if you want the company to pay in 30 days.

Your supplier may give you 30 days to pay, and you usually pay the bill in 20 days – you can change some payments around to make that work for you. Say you owe a certain supplier £30,000 a month and have an outstanding loan with an interest rate of 8.5 per cent. If you pay your supplier in the required 30 days (rather than 20 days), you can pay off your loan 10 days earlier each month, thereby reducing your total interest payments. You save £850 a year in interest payments and come out ahead of the game.

You can also try to get people who owe you money to pay sooner and improve your cash flow in both directions. Your financial adviser can help you determine what makes sense in your situation.

Letting employees invest

Employees who show their faith in the company by agreeing to work there may also have enough faith to invest in the company. Setting up a programme whereby your employees contribute a portion of their salary for shares in the company can improve the company's financial situation because you don't have to pay the employee the amount he invests, and your employees get a greater stake in making the company successful. If you do a good job your employees don't need much convincing that a pound they invest in the company may be worth a lot more than a pound in the future.

Venturing over to venture capitalists

Venture capitalists are institutional investors who provide start-up capital to high risk, high return, start-up businesses. A venture capital firm may be a group of wealthy individuals, a government-funded or -assisted organisation, or a major financial institution. Most venture capital companies specialise in a few closely related industries, such as medical products or energy-related technologies. If you have an invention that's going to revolutionise its field, venture capitalists want to fund it.

The right venture capitalist can make the difference between failure and success, growth and major growth.

Approaching a venture capital firm

No one finds approaching a venture capital firm easy. You have to seek them out; they don't come looking for you. For a comprehensive listing of venture capital firms that operate in the United Kingdom, visit the British Private Equity and Venture Capital Association Web site at www.bvca.co.uk.

Try to approach a venture capitalist before you need capital, not at the last minute. If you don't need the funding immediately, you're in a strong negotiating position. If you need money to pay staff, however, the venture capitalist has more power and can demand more concessions. Remember, the more you need the financing, the more it costs your company and management in the amount of control you have to give up.

Knowing what to expect

Various venture capitalists have different approaches to managing the business they fund and invest in. They generally prefer to take a very passive role in the day-to-day operations of a business; but if the business doesn't perform as expected, they may take an immediate and active role, insisting on changes in management or strategy.

Venture capitalists bank on the possibility that your company may go public and start trading on a stock exchange. If you envisage a family-owned company, don't look for funding from this source.

You may see a few clauses in an agreement with a venture capitalist that you don't see in contracts with other types of funders, including the following:

- ✔ A large interest in the company, sometimes more than half.
- ✔ A substantially higher rate of *return on investment* (ROI), which is the ratio of money gained on the investment to the amount invested.
- ✔ The right to assume control of the company in certain situations – if the company's financial position plummets, for example.
- ✔ The option of buyback rights for you so that you can regain full control of your company.

Always include a buyback option in your negotiations upfront. That way, when you pay back the original money you borrowed, you have an option to take back control of your company. Here's an example of how it works: When a well-known restaurant was getting started, the entrepreneur needed £1 million. He had £400,000 of his own money, so he needed £600,000 additional funding. Twenty investors put in £30,000, and each investor owned 3 per cent of the company. The investors owned a total of 60 per cent of the company and the entrepreneur only 40 per cent. When the entrepreneur paid back the investors their initial investment for their high risk, each of the investors then still owned 2 per cent of the company, thus in total they owned 40 per cent, rather than 60 per cent. The entrepreneur regained control of his company.

Looking at the pros and cons

Venture capitalists have a lot of money and a lot of contacts that can aid you in a lot of ways:

- ✔ They can jump-start product development and the regulatory process.
- ✔ They may have established relationships with production and manufacturing facilities that can make your production process less expensive.
- ✔ They've established an effective business system you can adopt immediately.
- ✔ Their ranks may include experts in your field who can assist and advise you. (Probably at least one member of the venture capital group sits on your company's board of directors.)
- ✔ They may run an *incubator* – a place where a lot of start-ups come together to take advantage of economies of scale by sharing common overhead costs, such as rent, a receptionist, legal and accounting services, Internet connections, office machines, and so on.

All the money, expert advice, and experience venture capitalists offer do come at a price, though:

- ✔ They typically want a significant portion of the ownership in your company – often the majority stake. When someone else owns 50 per cent or more of your company, you no longer work for yourself (thus, the importance of the buyback option – see the preceding section).

  Sometimes, a venture capital group creates a structure whereby it assumes majority ownership only if your company or product doesn't get off the ground or if things go badly.

- ✔ They almost always want at least some representation on your board of directors. Make sure that you can live with the representative who may show up at your office offering advice much more often than you may like.

- ✔ They may pull out if the company doesn't meet certain milestones. Sometimes, the venture capitalists phase in the investment over time based on predetermined milestones, and if the company doesn't reach a milestone in the allotted period, the venture capital group may stop funding you.

- ✔ They may force a sale or merger in order to reap the profits they expect.

You must weigh the pros and cons against your opportunities to obtain financing from other sources. Don't forget to consider the non-cash contributions venture capitalists can make that end up saving or making you more money.

Applying for Grant Funding

The UK Government encourages enterprise and innovation by operating a number of grant-funding initiatives designed to help small businesses and thereby help the economy. The various local authorities and regional agencies sponsor grant programmes to support the development and commercialisation of new ideas, products, and technologies.

Grappling with grant basics

A *grant* offers you free money. A grant is not a loan, not a trade for a share of your business, not anything but free. Sounds good, doesn't it? But just because a grant offers you free money doesn't mean you can just sit back and let the pounds roll in. You have to apply, meet the rules and qualifications, and get selected for a grant.

Don't let the thought of a little paperwork put you off, though. Applying for money from a bank, private investor, or venture capitalist requires a substantial amount of paperwork too. And if you get a grant, you don't have to pay the money back or give up any ownership in your company.

You compete against other applicants for grant funding, so being successfully awarded a research or development grant isn't going to be a walk in the park! You have to prove that your business is more in need of grant funding than those of other applicants, so consider researching and contacting companies that have recently received similar funding and asking them for tips.

You may also have to abide by strict conditions that the grant provider sets. Any failure to follow these terms may result in you having to repay the grant in its entirety within a set period of time.

Sourcing grants

The sorts of grants available to you vary according to where you're based, and what sort of project you're undertaking. You may also find that there are particular grants available dependent on how old you are or whether your invention involves a contribution to green technology.

Grants available in England

The Department for Business, Enterprise and Regulatory Reform (BERR) operates a research and development grant funding system to help individuals and small- to medium-sized businesses in England. The regional development agencies manage applications. For more information, visit www.england srdas.com.

Several types of research and development grant are available, each targeted towards different types of enterprise. BERR offers the most relevant grants to inventors on a 'matched-funding' basis, which means that you need to provide part of the finance yourself to qualify for the grant, for example:

- ✔ **Micro project:** This type of grant covers the development of a low-cost prototype for an innovative product (or process) before any significant commercial exploitation takes place. The grant rate is 50 per cent, which means that you need to find 50 per cent of the project costs yourself. The level of support is up to £20,000 and these grants are only available to micro firms with less than ten employees.

- ✔ **Research project:** This grant is designed to part-finance a well-planned feasibility study into the commercial potential of turning innovative technology into a new scientific product or process. The study may include expert assessment, testing, data gathering, and the creation of a professional report. The basic grant rate is 60 per cent and the level of support is for between £20,000 to £75,000. You can receive 20 per cent of the grant at the start of the project. Research project grants are available to small firms with less than 50 employees.

Your local Business Link is one of the best places to get help and advice on applying for a research and development grant. Business Link can also provide information on other types of funding that may be available to you. For contact details visit the Business Link Web site at www.businesslink.gov.uk.

Grants in other parts of the UK

Scotland, Wales, and Northern Ireland operate similar schemes to research and development grants, known as Smart Awards. For more information about these and to find out whether you may be eligible, visit the business sections of the following Web sites:

✔ Scotland: www.scotland.gov.uk

✔ Wales: www.wales.gov.uk

✔ Northern Ireland: www.investni.com

More specific types of grant

Other grant-funding schemes are in place to encourage specific types of innovation. For example, the Carbon Trust supports the development of new technology that can help to reduce UK carbon dioxide emissions. The Carbon Trust's Applied Research Grant scheme enables you to apply online for up to £250,000 with a minimum 40 per cent match-funding requirement. For more information visit www.carbontrust.co.uk.

If you have a good business idea and are aged between 18 and 30, you may consider contacting the Prince's Trust. It's an organisation that provides start-up support through the provision of low-interest loans, grants, and advice from volunteer business mentors. Visit the Prince's Trust Web site (www.princes-trust.org.uk) to find out whether you're eligible to apply.

Chapter 16

Keeping Control with Your Own Business

*B*eing your own boss and having the creative freedom to follow your goals to financial independence is many people's dream. Small businesses account for a significant part of the economy and have a crucial role in driving its growth.

However, just because you *can* start your own business doesn't necessarily mean that you *should*. Being an entrepreneur is a tough job, and one you must consider carefully before taking on.

This chapter tells you how to plan your business structure to handle all the elements of making and selling of your invention.

Considering Carefully Before Taking the Plunge

To start a business to sell your invention, you need a whole range of skills that very few people have. As an inventor, you're certainly creative, and if you're ready to market your invention, you're persistent and determined. But are you also a financial whizz, a marketing genius, a winning salesperson, an efficient production manager, an empathic human resources manager, and a reliable truck driver all rolled into one? Of course, owning and operating your own business can be incredibly rewarding, but you need to be as sure as possible ahead of time that starting a business is the right decision for you.

The following sections help you to consider whether you have what it takes to be your own boss.

Asking yourself some basic questions

Entrepreneur means many things to many people, but you can be sure of a few things that being your own boss does not mean. You don't have a safety net, a retirement plan, sick leave, or paid holidays. You don't go home at 5 p.m. every day and leave work on your desk. You don't have a stress-free life in which you never get a call in the middle of the night with some business crisis. We don't intend to discourage you from starting to build your empire, but first off, you may want to consider some basic questions:

- **How deep is your commitment?** Are you willing to put in 60 to 80 hours a week or even more? Especially in the early years, small business owners often eat, sleep, and breathe their businesses. You may not get to take a holiday, and calling in sick won't achieve much either! Do you have enough enthusiasm and commitment to your invention, and to every aspect of producing and selling it, that devoting your life to making it happen sounds like fun? If you're cringing just reading this, business ownership isn't for you.

- **Do you have the support of your partner and family?** Aside from your commitment, your partner and other family members have to share some level of commitment to your business. You'll dedicate your energies to the business and won't have as much time or energy for your family as you'd like to have.

You may also put your family in a risky financial situation. You may all have to adjust to a lower standard of living and do without exotic holidays, new cars, or possibly even a Friday night drink at the local pub! Think long and hard about the personal and social implications that starting a business has for you and your family. The compromises won't necessarily be short term either. Are you the sort of person who's happy not to keep up with the Joneses?

- **Is your knowledge base broad enough?** If your business is going to design, manufacture, sell, and distribute your product, you need to know more than a little bit about all aspects of the process. As the head honcho, it's up to you to troubleshoot when the production line hits a snag, transportation gets waylaid, or any other problem crops up. If you don't have a clue about mechanical things, your production line snag may become a major headache. Likewise, without some familiarity of distribution systems, your transportation trouble may run off the rails.

Of course, you gain knowledge as you go, but you need to have basic familiarity to get started.

Now you've pondered these questions, things start to make sense, or you begin to have doubts. Either way, listen to your feelings and trust yourself.

Evaluating your skills

Start out by evaluating your strengths and weaknesses as the owner and manager of a small business. Sit down and ask yourself the following questions:

- **Am I a self-starter?** You'll be entirely responsible for developing projects, organising your time, and following through on details. If someone has to pounce on you to get you out of bed in the morning, think twice before starting your new adventure.

- **How well do I get along with different personalities?** Realise that business owners typically need to develop working relationships with a variety of people, including customers, vendors, buyers, staff, bankers, and professionals such as lawyers, accountants, and consultants. Can you deal with a demanding customer, an unreliable vendor, or a cranky receptionist and maintain a productive relationship? You may be a people person, but be aware that you'll encounter every mood and attitude.

- **Am I good at making decisions?** You have to make decisions constantly as a small business owner – some with large consequences and some not so large; some in an instant and some after taking time to explore options. You may have an adviser to bounce things off of, but you're the final authority and must make many decisions all by yourself.

- **Do I have enough physical and emotional stamina?** Running a business can wear you down. You put in long hours six or seven days a week. You're not home enough to spend quality time with your family. You have commitments to vendors, customers, and employees and responsibilities to meet here, there, and everywhere – the weight is on your shoulders. You need strong motivation to help you survive the ups and downs.

- **Am I good at planning and organising?** Many business fail as the result of poor planning. Good organisational systems throughout the company improve efficiency and results, and help make planning easier. How good are you at making plans and setting up systems?

To start your own business you need support, but you also need to have some basic resources within yourself if your business has a chance of success.

Realising that most small businesses fail

Success is never automatic. Business success depends primarily on your foresight and organisation – and a little luck never hurts. But even then, you've no guarantees.

Over 50% of small businesses fail in the first year and 95% fail within the first five years. To give yourself and your business the best chance of being one of the successes, try to avoid the top reasons for failure:

- **Poor management:** Lack of experience, faulty systems such as inventory control.

- **Money troubles:** Insufficient capital, inefficient use of capital, poor control of cash flow, disadvantageous credit arrangements, over-investment in fixed assets, use of business credit for personal use.

- **Market issues:** Strong competition, out-of-control growth.

- **Product issues:** Lack of intellectual property protection or lack of patent defence.

Remember that underestimating the difficulty of starting a business is one of the biggest traps for inventors and entrepreneurs. Don't overestimate your chances of success. But if you're patient, willing to work hard, and prepared to take all the necessary steps and be receptive to constructive advice, then you improve your chances of success.

Deciding to Go for It

Entrepreneurs make up the backbone of a country's economic growth. They're self-motivated and make things happen. Large businesses may get most of the attention, but small businesses drive the economy.

A new venture is risky, but exciting. Many reasons exist not to start your own business. But for the right person, the advantages of business ownership far outweigh the risks:

- You get to be your own boss.

- Your hard work directly benefits you, rather than increasing profits for someone else.

- Your earning and growth potential have no limits.

Running a business provides endless variety, challenges, and opportunities to grow.

In business, just as in life, you don't get guarantees. You simply can't eliminate all the risks associated with starting a small business – but you can improve your chances of success with preparation, insight, and good planning.

Assessing your skills, education, and experience

Your skills, education, and experience provide you with tools that you can use in starting your own business. Recognising what you have helps you figure out what abilities you need to develop in order to give your business its best shot at success. To assess what skills you already possess, follow these steps:

1. **Sit down with some paper and a pen.**

   You may need a few sheets of paper.

2. **List your personal characteristics.**

   List your ten main strengths and five to seven serious weaknesses. Get someone close to you to do the same thing, and then discuss the similarities and differences, and situations in which you used your talents.

   Give yourself extra points if you list any of the following (the characteristics that successful entrepreneurs share): competitive, creative, demanding, a desire for immediate feedback whether positive or negative, goal-oriented, high energy levels, independent, innovative, inquisitive, persistent, risk-taking, self-motivated, self-confident, and a strong drive to achieve.

3. **Make a short detailed list, not more than two pages, of what you know about the field or industry you're getting into.**

   If you're well versed, simply outline the general areas. If you're not sure, list everything. Concentrate on specific business information.

4. **Identify your entrepreneurial skills.**

   If you start your own business, your innovative abilities don't stop with simply having a new product. If you're a very laid-back couch potato who doesn't like to travel, hates being a salesperson, and relies primarily on others to make decisions, think twice before starting your own business. You may want to consider licensing as an option (see Chapter 21).

5. **Analyse two recent decisions you made in your current job that did not work out well.**

   Figure out why things didn't work and try to identify the point in the process where a different decision may have made a difference.

6. **Write down the particulars of two recent problems that you solved in your job successfully.**

   Analyse the processes and look for ways to incorporate the steps of those good decisions into everyday decision making.

Taking a clear look at your source of motivation and your belief in yourself helps you to prepare for the challenges that come with owning your own business.

You probably don't have all the skills and experience you need to start-up and scale a new business successfully. If your skills-set falls short in a certain area, why not consider creating a board of advisers? This isn't a board of directors – which is a legal entity whose members may, in fact, have useful advice to offer – but a group of knowledgeable people with a wide variety of interests and expertise.

You can't do it alone. Surround yourself with the best that you possibly can and then take the most important step and *listen to them*.

Planning the steps

Like a chess game, success in small business starts with decisive and correct opening moves. And although initial mistakes aren't fatal, it takes skill, discipline, and hard work to regain the advantage.

To increase your chance for success, take the time upfront to explore and evaluate your business and personal goals. Then use this information to build a comprehensive and well-thought-out business plan to help you reach these goals.

Follow these steps to get your business off the ground:

1. **Apply for a patent or other intellectual property rights.**

   If a patent search shows an existing patent exists for your product (which is highly likely), you may want to nip your business venture in the bud. Chapter 4 explains how you search for intellectual property rights.

2. **Conduct a market feasibility study.**

   Make sure that you have a market for your product before spending a lot of time and money. If feedback on your invention is negative and consumers don't want to purchase the product or if the manufacturing cost is too high for what consumers are willing to pay, drop the idea. Stay at home or you'll lose money. Flick to Chapter 18 for details on how to conduct market research.

3. **Write a business plan.**

   The process of developing a business plan helps you think through some important issues that you may not have considered yet. Your plan becomes a valuable tool as you set out to raise money for your business. It also provides milestones that you can use to gauge your success. Chapter 14 guides you through your writing a business plan.

4. **Fix funding.**

   Securing funds for your venture is critical. We dedicate Chapter 15 to guiding you through your funding options.

5. **Nail down the location(s).**

   If your manufacturing plant and your sales shop are in separate locations, you need to figure out distribution channels as well. Insufficient attention to distribution can be a major source of problems for business start-up and growth. Read more on choosing a location in the 'Finding a home for your business' section, later in this chapter.

6. **Get the manufacturing plant operational.**

   Whether you plan to manufacture in your own plant or through an established manufacturer, pay particular attention to the necessary steps for start-up. Chapter 17 discusses manufacturing arrangements.

7. **Sign agreements with suppliers, vendors, and distributors.**

   Negotiate and sign agreements at the earliest possible stage with the third parties you'll be working with to get the product to market. Check out Chapter 17 for details of partnership arrangements.

8. **Take out necessary insurance policies.**

   What if the factory or warehouse is burgled, burns down, or gets hit by a hurricane or flood? Put insurance in place as soon as you can. Get the basics in the 'Taking out business insurance' section, later in this chapter.

9. **Design a marketing plan.**

   Plan in advance your route to market and how you intend it to grow. Chapter 19 has the low-down on marketing your product.

10. **Launch your sales campaign.**

    Think about a style of launch that suits your product, from targeted letters to the potential buyers of a specialist item to a glitzy celebrity launch with flashing lights, waterfalls, and fireworks. Skip to Chapter 19 for more ideas on your launch.

11. **Hire employees.**

    Getting the right employees is one of the most vital jobs, and one that is notoriously difficult to do. Don't be afraid of using professional recruiters, but choose them well. See Chapter 11 for information about hiring people to make your business a success.

12. **Open your doors.**

    Ready, steady, *go!*

Along the way, you've many questions to answer and decisions to make. You need to consider what advantage you have over existing firms and whether you can deliver a better quality service, and then you must create a demand for your business. And don't forget such seemingly mundane matters as what to name your business and how to pay yourself.

Finding a home for your business

If you take on the manufacturing, marketing, and selling of your invention, you need space to accomplish all those tasks. And you need different types of spaces depending on whether you plan to do everything in one place or in two or more different locations.

Working out how much space you need

On the face of it, it makes sense to do all your manufacturing and selling in the same place, but the nature of the manufacturing process may make it more desirable to assemble your invention in one place and sell it in another. Of course, if you sell by mail or online and don't need a shop on the high street, you can probably dispatch directly from the manufacturing plant and not worry if the production process is loud or smelly.

You may even have more than one manufacturing location, and tool a proportion of your product in one place and then transfer it somewhere else for the finishing touches.

You have to take all sorts of things into consideration when looking for space. If you sell your invention from a shop, the old saying 'location, location, location' comes into play. Your market research (see Chapters 18 and 19 for details) tells you the best area for your business. You then have to find a place you can afford and that meets your other requirements.

If you don't need a shop front, distribution becomes your main concern. You want to be conveniently located for your preferred distribution method, whether that's land, sea, rail, or air, and keep alternative transportation in mind in case of need.

Don't forget to consider storage for your raw materials and finished product. It may be cheaper to lease a storage space than to lease a space big enough to house both your manufacturing or sales floor and your stored goods.

Leasing or buying

Whether you decide to lease or buy a building to operate your business from, you spend a lot of money. For a new business leasing is the less expensive option: You don't pay for the big-ticket items. If you buy, you need to be prepared to pay very high start-up costs for the purchase price, equipment, and renovations, and many other expenses. Investing in a manufacturing plant moves your break-even point way out into the future – a future that you don't know that your fledgling company will ever see. In the worst case scenario, you're better off buying out the rest of a relatively short-term lease than owing 30 years' worth of payments for a building and equipment you no longer use.

If you decide to lease space to operate your business, you need to ask some of the following questions before signing the lease agreement:

- Is the tenant's share of expenses based on total square footage of the building or the square footage leased by the landlord? Your share may be lower if it's based on the total square footage.

- If the tenant stops using the building, does the lease define the remedies available to the tenant, such as rent abatement or lease cancellation?

- If the landlord doesn't meet repair responsibilities, can the tenant make the repairs, after notice to the landlord, and deduct the cost from the rent?

- Does the lease clearly define how parties can settle disputes?

Speaking the language of leases

You need to understand the terms used in your lease. Some of the most common ones are:

- **Anchor tenant:** This is a department store or supermarket that attracts customers to a shopping centre.

- **Assignment:** Here the tenant turns the lease over to another business, that assumes payments and obligations under the lease.

- **Common areas:** Maintenance charges sometimes apply to 'common areas', which several tenants may use. Such charges include property taxes, security, car park lighting, and maintenance.

- **Exclusivity provision:** For example, a shopping centre can't lease to another company that provides the same product or service that an existing tenant does.

- **Gross lease:** The tenant pays a flat monthly amount and the landlord pays all operating costs, including property taxes, insurance, and utilities.

- **Lessee:** This is another way to say tenant.

- **Lessor:** This is a fancy word for landlord.

- **Percentage lease:** This includes base rent, operating expenses, common area maintenance, plus a percentage of the tenant's gross income. A percentage lease arrangement is most common for retail leases.

- **Right of first refusal:** Before renting vacant space to someone else, the landlord must offer it to the current tenant with the same terms that the landlord offers to the public.

- **Sublease:** The tenant rents all or part of space to another business, but the tenant is still responsible for paying all costs to landlord.

Taking out business insurance

Like home insurance, business insurance protects the contents of your business against fire, theft, and other losses. All businesses must purchase certain basic types of insurance. The law requires some types of coverage; others simply make good business sense. The types of insurance that we list below are among the most commonly used, and which you can use as a starting point for evaluating the needs of your business:

- ✔ **Liability insurance:** Your business may incur various forms of liability, chief among them *product liability insurance* (PLI). Under the 1987 Consumer Protection Act, you may be legally responsible if your invention causes damage or injury. PLI protects you against the compensation a court awards if your invention harms a customer or property. Most PLI policies provide cover of about £2 million. You may have other types of liability related to your specific industry. Liability insurance covers you in the event that someone makes a liability claim against your business.

  Liability law is constantly changing and is complicated enough even when it doesn't change. Getting a competent professional to assess your liability insurance needs is vital in determining an adequate and appropriate level of protection for your business. Implementing measures such as comprehensive quality control can help to reduce your insurance premiums.

- ✔ **Property insurance:** If you're a tenant, check what insurance your landlord already provides for the property. If you own the property, many different types of property insurance and levels of coverage are available. You need to determine how much of your property you need to insure for the continuation of your business, and the level of insurance you need to replace or rebuild. You must also understand the terms of the insurance – including what the insurance doesn't cover.

- ✔ **Contents insurance:** Your property insurance covers the buildings in which your business operates. You must separately insure whatever is inside those buildings. Contents includes your raw materials, finished stock, work-in-progress, machinery, furniture, and other contents.

- ✔ **Key person insurance:** If you (and/or any other individual) are so critical to the operation of your business that it can't continue in the event of your illness or death, consider key person insurance. The funds can come in handy during a period of ownership transition caused by the death or incapacitation of a key employee.

- ✔ **Motor insurance:** Obviously, you need to insure any vehicle that your business owns for both liability and replacement purposes. If you have a fleet of delivery vehicles then consider a flexible fleet insurance policy to reduce overall cost. Also find out what restrictions exist on the policy. For example, is an 18-year-old driver with nine points on her licence insured to drive your shiny new delivery van?

✔ **Intellectual property insurance:** If another company infringes your intellectual property rights (such as a patent or registered design), you may suffer a loss of revenue and market share. Intellectual property insurance can provide some financial help to take legal action against the infringer (see also Chapter 6).

✔ **Insurance on a home office:** If you establish an office in your home, contact your insurance provider to update your policy to include coverage for office equipment. Home office coverage isn't automatically included in a standard homeowner's policy.

These more important types of insurance are just a sampling of those you need to consider. Consult a qualified insurance agent for advice.

Building a Business Structure

When organising a new business, one of the most important decisions you make is choosing your business structure. Your decisions have long-term implications, so consult with business advisers, including your accountant and solicitor, in order to select the form of ownership that's right for you.

For more help with planning the structure for your business, consult the following organisations:

✔ Business Eye, Wales. www.businesseye.org.uk.

✔ Business Gateway, Scotland. www.bgateway.com.

✔ Business Link, England. www.businesslink.gov.uk.

✔ Highlands & Islands Enterprise, Scotland. www.hiebusiness.co.uk.

✔ InvestNI, Northern Ireland. www.investni.com.

Going it alone as a sole trader

Becoming a sole trader is the easiest and least costly way of starting a business. You control the business and assume the day-to-day responsibility of running the operation and making key decisions.

Being a sole trader has several advantages, including the following:

✔ You're in complete control and you can make decisions as you see fit.

✔ The business structure and hierarchy is as simple, or as complicated, as you make it.

✔ You receive all the income the business generates and you choose whether to keep it or reinvest it.

✔ The business is easy to dissolve. Selling a sole-trader business can be as simple as selling a house.

With good comes bad, and several disadvantages attach to being a sole trader:

✔ You have unlimited liability and are legally responsible for all the business's debts. Your business and personal assets are at risk.

✔ You may be at a disadvantage in fundraising. Venture capitalists, for example, require a share of the business in return for their investment (see Chapter 15 for the low-down on venture capitalists). You may be limited to using your personal savings or taking out personal loans.

✔ You may have a hard time attracting high-calibre employees, because sole-trader businesses generally pay less and have less room for advancement. You may have to give employees an opportunity to own a part of the business.

To become a sole trader you need to register as self-employed with HM Revenue and Customs. For more information, visit www.hmrc.gov.uk.

Partnering up

The two most common types of partnerships are *general* and *limited*. A *general partnership* is simply an agreement between two or more persons to enter into business together. The agreement can be oral, but we strongly recommend that you draw up a professional agreement for all parties to sign. Consult an accountant or solicitor if you need help with this.

You can also set up a *limited liability partnership*, which allows several investors to create a common business relationship while limiting each investor's personal liability to the amount that the investor put in.

A partnership agreement must be very clear about all partners' duties and responsibilities. As well as detailing the type of business, the agreement covers the following areas:

✔ **Control:** Who makes the decisions on a day-to-day basis; the process for making long-term business decisions; provisions for changing the partnership in general or due to death or incapacitation of a partner; processes for settling disputes among partners.

> ✔ **Money:** The amount each partner invested; how you distribute profits (or losses) among partners, both ongoing and when the partnership dissolves; provisions limiting expenditures under certain conditions or by certain people.
>
> ✔ **Time:** Amount of time each partner contributes; how long the partnership is in effect.

Make sure that your agreement is specific about terms and conditions for partner buyouts. You probably don't want to think about breaking up when your partnership is so new, but many partnerships split up during times of crisis. Unless you've set up a defined process, you may be in for even greater problems. In a general partnership, selling or transferring an individual partner's share to someone else may require agreement from all partners. If a partner dies, the partnership dissolves, so have a contingency plan for that as well.

As a partner, you retain significant control of your business and can make the business as big or small as you and your partners like. Partnerships can go along swimmingly for years and make lots of money for all parties involved. However, all partners are responsible for each other's business actions, which may land you in hot water if one of your partners makes a poor decision.

The partnership itself isn't responsible for paying taxes on the income that the business generates. The partnership files a tax return for informational purposes only. Each partner pays taxes on her share of the business income. The profits and losses from the partnership flow directly to the individual partners.

Looking at private limited companies

In England, Wales, and Scotland you register private limited companies at Companies House (www.companieshouse.gov.uk). In Northern Ireland, you register these types of businesses with the Companies Registry (www.detini.gov.uk).

You can set up a private limited company (limited company, for short) from scratch or purchase one off-the-shelf. In either case, you must appoint at least two people: a director, who must act in the best interests of the business, and a company secretary. (But from October 2008 it's no longer necessary for a private limited company to appoint a company secretary.) The company normally employs its staff through the PAYE (Pay As You Earn) tax system. Consult with your accountant to find out more about PAYE.

Shareholders of a private limited company can be individuals or other companies. Unless the company guarantees some sort of security (such as a personal asset) the shareholders aren't responsible for outstanding company debts if the business collapses. If the company fails the shareholders lose their investments, but not necessarily the shirts off their backs!

A limited company name is different from a registered trade mark. Turn to Chapter 8 to find out what a trade mark is and how to register one.

Private limited companies must:

- Never offer or issue shares to the general public.
- Keep accounting records and file them annually with Companies House (for businesses in England, Wales and Scotland) or the Companies Registry (for businesses in Northern Ireland).
- Pay corporation tax on profits earned.
- Raise finance from shareholders, by borrowing from established lenders, or from retained profits.
- Have a registered address within the UK, which is also a valid postal address.

Seek the advice and services of a qualified accountant before setting up your own limited company. The accountant can guide you through the company registration process, advise you on what records you need to keep, and manage the preparation of annual accounts. Accountants can be worth their weight in gold (literally) when it comes to calculating tax liability.

When choosing an accountant don't just go for the first company you find listed in the *Yellow Pages*! Ask friends and family for recommendations and ask the accountant for references and a summary of fees before signing up.

Considering public limited companies (plcs)

A public limited company (plc) is a legal entity, owned by the shareholders . The law considers the plc to be a unique entity, separate and apart from those who own it. Public limited companies pay tax, can be sued, and can enter into contractual agreements. Basically, plcs have a life of their own and don't dissolve when ownership changes. The owners/shareholders aren't liable for company debts at any level.

The business owners are referred to as *shareholders* – individuals or companies that hold shares in the organisation. The shareholders elect a board of directors to oversee the major policies and decisions, which a management team executes. Regular board of directors' meetings and annual shareholders' meetings shape those decisions.

The control of a public limited company is based on share ownership. The person (or people) with the largest number of shares controls the company. Whoever effectively controls 51 per cent of the shares is able to make policy decisions, whether that's a group of shareholders acting together or an individual. And unlike a private limited company, a plc can offer its shares for sale to the general public through a stock exchange.

The advantages of forming a public limited company are:

✔ As a shareholder, you have limited liability for the business's debts or for legal judgments against the company.　Generally, shareholders can only be held accountable for their investment in shares of the company.

✔ Investors may be eager to trade their money for shares in the company, and if the company needs to raise cash, it can generally sell shares quickly and easily.

Becoming a public limited company and selling shares in your company means sacrificing significant control. You don't make decisions; the board of directors does. The legal and tax ramifications of a plc are fairly complicated – as is everything else about this most complicated of all business structures.

Managing Your Business the Right Way

Setting up a company is just the first hurdle in establishing a successful long-term business. You need to plan and manage a whole range of other factors, including:

✔ **Employment:** You need to be aware of, and comply with, minimum wage legislation and establish a clear policy towards the provision of pension plans, holiday allowance, sick pay, equal opportunities, and disciplinary procedures for your staff.

✔ **Health and safety:** The law obliges you to ensure that the working environment is safe in order to protect the welfare of your staff and customers. You must have the necessary insurances to cover against accidents and the unexpected.

✔ **The environment:** You need to think about how your business affects the environment. Are there regulations for disposing of your potentially hazardous waste materials? How are you to control pollution that your factory generates? Would gaining accreditation for your high environmental operating standards (such as ISO 14000) be beneficial to your business?

You're not alone! For up-to-date information and help with company formation, trading regulations, government legislation, and fulfilling your obligations as an employer, visit the Business Link Web site at www.businesslink.gov.uk.

Chapter 17

Partnering and Manufacturing Arrangements

*U*nless you're an inventor of independent means – and even if you are – you probably aren't prepared to buy a factory to produce your invention. This chapter gives you pointers on making arrangements to produce your product in some kind of partnership, or on a contractual basis.

You need price estimates for making your product and you must decide whether to make it yourself or locate an existing manufacturing company that can make it for you in your home country, or offshore. And if you're wondering how to manufacture your product on limited funds then you may consider a partnership with another company.

Just because *you* think that your invention is the best thing since sliced bread, it doesn't necessarily mean that everyone wants to buy it. Don't get caught out by spending large amounts of money on setting up production and manufacturing your invention in large numbers without having adequate orders in place to fulfil.

Paying and Partnering Arrangements

If you're a lone inventor and don't have any experience in manufacturing, working with an existing manufacturing company is wise. Doing so saves you time and money in the long run.

You've a variety of options for producing your invention: The most relevant may be to go it alone and pay a manufacturer to make your product (called *subcontracting*), or to persuade a manufacturer to bear some of the production costs (an arrangement called a *joint venture*). The following sections explore these options in greater detail.

Subcontracting

Subcontracting means paying a manufacturer to produce your product. If you subcontract, you pay all the production expenses according to the terms you work out with the manufacturer. If your product sells slowly, you don't get more time to pay; if the distributor loses a shipment, you don't get more time to pay; and so on.

To make a subcontracting arrangement cost effective, you need to make the order pretty large. And to make producing your invention worthwhile for the manufacturer, you usually have to pay part, or all, the cost upfront.

You know for certain that your invention is the wonder of the ages and is going to sell like hot cakes the minute it hits the market. Unfortunately, most manufacturers run into a dozen inventors just like you and may get burned by a few of those inventors whose products don't sell and who end up not paying their bills. The decision-maker at the manufacturer has to answer to shareholders and can't afford to believe in your dream as much as you do. He has to believe in cash upfront.

Be cautious when working with manufacturers. Have a patent or strong confidentiality agreement (more on these in Chapter 2). Remember, if a manufacturing company can make the product for you, it can make the product for itself. Often, a manufacturer also has a distribution network in place. Manufacturers have been known to work around a patent and come out with their own version, so make sure that you're thoroughly protected.

Going into a joint venture

A *joint venture*, as we use the term in this context, is an agreement between two or more interested parties to produce and market a product or a service. Essentially, a joint venture spreads the costs and risks between more than one individual or company.

If you don't have money to pay an existing manufacturer to produce the tooling and create mass quantities of your invention, consider trying a joint venture with an existing company in order to share some of the costs involved.

If a manufacturing company believes that your invention is going to fly off the shelves, it may be interested in a joint venture, envisaging massive reorders and lots of profit.

Don't confuse *joint* with *equal*. Generally, you pay the majority of the production costs, often upfront. The manufacturing company may only be willing to pay for making the production tooling you need, or may possibly extend you some credit beyond its normal terms.

Teaming Up with a Manufacturer

If you don't have the time or know-how to manufacture your product yourself, you need to find someone who does. Whether you license your invention to a company that plans to manufacture, market, and sell it (see Chapters 21 and 22 for more on licensing) or decide to bring your product to market yourself, you still need to speak with manufacturing companies.

Whether you license your product or not, the more information you have regarding the costs for production, the better off you are. Without manufacturing estimates for production, you can't determine profit margins. So if you want to license your invention, speak to manufacturers to get production cost projections so that you can bring that information to the table when trying to get a licensing agreement.

Hooking up with a manufacturer resembles a marriage. Make sure that you're compatible and can live comfortably together for the long haul. Ensure that you get production quotes in writing and go through agreements with a fine-tooth comb.

Locating potential manufacturers

Manufacturing output in western countries has generally been in decline for many years. Companies that specialise in developing new consumer products increasingly turn to the Far East to fulfil their manufacturing needs (see the section 'Partnering Abroad', later in this chapter). However, if local production *is* commercially viable for your invention, you can draw upon many sources in your research to locate an appropriate manufacturer. These sources include:

✔ **Business Link (www.businesslink.gov.uk):** A government support service that provides impartial, practical advice. Most Business Links operate a useful supplier-matching service through which you can source contact details for approved local manufacturers.

✔ **Confederation of British Industry (CBI; `www.cbidirectory.co.uk`):** A useful online directory that includes a search facility for business names and categories.

✔ **Manufacturing Advisory Service (`www.mas.dti.gov.uk`):** A UK government organisation that provides information about different types of manufacturers, general advice, and fact sheets.

✔ **Trade directories:** Try researching company information in your local library. You can find the *Kelly Search* and *Kompass* directories in the business reference section as well as online at `www.kellysearch.co.uk` and `www.kompass.co.uk`.

Consider running a credit check on a manufacturing company you like, just to make sure that it's in good financial health. The company's beautiful new building may mask financial disaster within the walls. Speak to your local bank, which can run credit checks on your behalf.

You may also be able to gain information on UK companies from the UK Companies House 'WebCheck' service, which is available online. For a small fee you can download a company's latest accounts, annual returns, and certain company reports. You can search for a company by name or company registration number. Visit `www.companieshouse.gov.uk` to access the WebCheck database.

Assessing plants

Hiring a manufacturing firm to produce your invention is often the most economical way to get your product to market. You don't have to build a factory, hire engineers, lease the machines, or do any of the things you have to do if you manufacture your own product. Of course, even though you've a whole list of things you don't have to do, you've an even longer list of things you do have to consider. We list below some of the issues to consider and questions to ask about the facilities as you look for a manufacturer to make your product for you:

✔ **Number of years in business, reputation, and financial health:** An established company lends you credibility with distributors and retailers. Some retailers want to know which manufacturer is producing the product before they agree to carry it. To make their money, retailers have to keep their shelves stocked. If they know from experience that a certain manufacturer meets its delivery deadlines, the retailer is more likely to agree to stock a product made by that company.

✔ **Compatibility with your product:** If a manufacturer makes similar products to yours, you can have some confidence that it has the ability to make your product. You can also check out the quality of the manufacturer's work. Make sure that it has the proper equipment to make your product. If your invention uses metal tooling, does the manufacturer use aluminium or steel? Aluminium is generally cheaper, but steel tooling lasts longer, allowing you to make more parts from one tool. You want to use the method that best suits your needs. If you need metal turning or stamping, does the manufacturer have the experience and equipment to do that?

✔ **Age and size of facilities:** Does the factory boast the latest, most efficient and cost-effective machinery? Older machinery may break down, causing delays, missed delivery deadlines, broken contracts, and massive headaches for you.

✔ **Capacity:** Can the factory finish the manufacturing process in-house, or does it have to outsource any of the steps? If the manufacturer subcontracts, you need to investigate the subcontracting company as well. Does the factory have enough capacity to produce your invention? If the machines have to work round the clock to meet current needs, when can they make your product? If your product must be compatible with consumer purchasing habits, such as holiday and seasonal demand, can the factory be sufficiently flexible?

✔ **Standards:** If your product has to meet standards imposed by government agencies, can the factory meet those standards? How good is the factory's quality-control process? How does the company deal with defective items, and what kind of recompense can you expect if the factory ruins a whole production run?

✔ **Timeliness:** What happens when the factory gets backlogged? Become familiar with *penalty clauses*, which set out the remedial action the manufacturer takes in the event it can't meet your deadlines. A penalty clause makes the subcontractor financially responsible and accountable for your loss of sales, as well as the potential loss of your account with retailers that you sell your product to.

Suppliers can lose accounts with major retailers if they fail to deliver their product on time. If you miss a delivery date, you may get a second chance from a retailer if you're lucky – but don't expect a third. When your product isn't on the retailers' shelves on time, shelf space is empty and that causes the retailer to lose money. The burden is on you and the fault lies with your manufacturing company.

✔ **Costs and payment policies:** What is the estimated final cost in terms of tooling and production? What quantity discounts does the manufacturing company offer you for manufacturing your product? For example, what's the price per unit to have 1,000, 5,000, 10,000, or even 1 million units manufactured? Does its payment policy leave you any room to manoeuvre? Suppose that 10,000 units of your product ship directly to a major retailer. That retailer doesn't pay you for 60 days, but your payment to the manufacturer is due after 30 days. To add to the confusion, the retailer places another order for 50,000 units. Will your manufacturer extend you credit?

Product liability insurance is a very expensive necessity. Most of the larger retailers require you to have a product liability policy in place before they even consider selling your product. You may be able to save yourself money by having the manufacturer add your product onto its existing policy. The manufacturer should have liability coverage already (if it doesn't, you need to find out why), and the cost of adding your product to that policy is minimal – like adding a third child to a family insurance policy. If your manufacturer can't include you on its insurance, contact your own insurance provider. They may not handle product liability themselves, but they can probably locate a provider that does.

Examining the manufacturer's track record

Look at your manufacturing partner as a potential mate with whom you've a legally binding contract. Look at possible plants with a keen eye and an investigator's background (see the earlier section, 'Assessing plants', for details on how to find a good manufacturing partner). You need to work with a reputable company. If the manufacturer doesn't make your product on time or uses cheap materials, you won't receive repeat orders. A shoddy manufacturer can destroy you.

Speak with several different companies before making a decision. Don't just go to one manufacturing company and think that your job's done. Remember, you're going to pay a company to manufacture your product and that manufacturer gets paid, whether your product sells or not.

Your best bet is to check references. Talk to other inventors who use the manufacturing plant to produce their goods. Try to talk with someone that has the kind of continuing relationship you envisage your company having with a particular manufacturer.

Compare the results of the questions you ask, listing the pros and cons of the various manufacturing companies and the overall costs associated with the actual manufacturing of your new product. This comparison helps you narrow down your choices and may help you make your final decision.

Negotiating a contract

Making a deal with a potential manufacturer gives you one more opportunity to sell yourself and your product. You want the production people to be as enthusiastic as you are about your invention so that they want to do it right. You may also convince them to invest a little money too, by becoming a joint partner (see the 'Going into a joint venture' section, earlier in this chapter, for more on this type of arrangement).

Before you meet with potential manufacturers, put yourself in their shoes and ask yourself what terms and conditions you can live with, as a manufacturer. A good deal – a workable, lasting agreement – is one that all parties think treats them fairly.

You must realise that your manufacturer has to make a profit in order to stay in business, just as you do. The manufacturing environment constantly changes. Governmental regulations can change with the stroke of a pen and cost thousands of pounds in factory improvements; raw materials can double in price in days and triple in months; an act of God – such as flooding, earthquakes, hurricanes, and lightning – or power loss can impact the factory for hours, days, weeks, or months. Your chosen manufacturer certainly has these possibilities in mind as you negotiate the terms of your deal, so you need to be aware of them also.

Just as you're concerned about the integrity and reliability of the manufacturer, the manufacturer has those same concerns about you. You may consider yourself to be a completely honest, well-meaning, and well-intentioned individual, but the manufacturer doesn't know that, and an irresponsible inventor may have burned it in the past. Some of the issues a manufacturer may want to address include:

- **Your financial health:** If your product doesn't sell, or doesn't sell as quickly as you anticipate, can you still pay your bill?

- **Your product's prospects:** How good is your product? What have you done to determine whether your product has a market? Do you have retailers on board to sell your product? Does your invention have good patent protection? If not, the manufacturer is put in a liability situation for possibly infringing on someone else's patent and that person can sue the manufacturer.

- **The workability of your prototype:** Does your prototype easily lend itself to duplication and mass production? If not, who's responsible for making a working model or production tooling?

All too often, an inventor is so much in love with his invention that the practicalities of duplicating it in the real world just don't register. A manufacturing partner can't afford to take the same attitude.

Partnering Abroad

To go to market, you need to manufacture your product at a competitive price. You may want to look at your offshore pricing options and see if you can save money by making your invention offshore. You may want to do so, especially if you need several thousand units. When manufacturing offshore, having only a few made isn't cost-effective: You purchase by the container.

Producing your invention in a foreign company has both advantages and disadvantages. This section covers the important issues to consider before you make a decision.

Looking at legal issues

Legal control is most likely to be your first consideration when you think about manufacturing in a foreign land. In setting up your partnership arrangements you should consider, and if at all possible, define in writing such aspects as the laws (for example English law) to be used in interpreting any disputes, and the legal form (for example the English courts) in which such disputes are to be settled.

You may not have the same patent protection in other countries as you do in the United Kingdom and you may therefore be vulnerable to someone copying your invention elsewhere. If the copying takes place in another country, you may have a much harder time in the legal arena. See Chapter 3 for more information about intellectual property.

You may have to find a reliable and trustworthy attorney who's qualified to practise in the country in which the factory operates. You also have the added expenses of travel and time away from your business to consider.

A UK patent gives you the right to sue in the United Kingdom. It doesn't mean that you have patent protection in other countries. In fact, many times offshore companies see a product that sells well in the UK and decide to make and sell it in other countries throughout the world. Unless you're a manufacturing giant with a lot of purchasing power, offshore manufacturers can become your greatest competitors. Offshore manufacturers make many highly desirable products and sell these knock-offs for pennies compared with the patented product.

If you take the offshore route, you need to consider the customs laws in your own country. Basically, you're importing your own invention. You don't have to hire a customs broker, but many importers opt to do so for the convenience. *Customs brokers* take the burden of filling out paperwork and arranging customs clearance for imported goods. Import rules can be particularly complex, especially regarding a new product, so a broker may be the way to go. You're ultimately responsible for meeting customs regulations, but a customs broker can save you from making costly mistakes.

Find out more about importing and exporting, and finding freight forwarders and customs brokers, by checking out the British International Freight Association Web site, www.bifa.org.

Pricing labour costs

Hiring workers in foreign countries can offer huge savings in labour costs. Asia in general and China in particular boast workers with highly developed technological manufacturing skills.

Compared with China, manufacturing labour costs in western Europe are enormous. For example, a factory may pay a highly skilled production tool designer in the UK anything from £30 to £50 per hour. The same skill in China may earn the worker about £1 per hour. So it's not surprising that so many manufactured products have the words *Made in China* stamped on them.

A worker in a developing country who has little or no technical skills makes under £0.50 an hour. Such a worker can competently cut and sew garments or work on an assembly line.

The cheap labour costs may be an irresistible lure, but Western society is becoming more socially conscious. If someone accuses your foreign plant of being a sweatshop that exploits the local workers without adequate compensation, your reputation and sales may suffer.

Probing product cost and quality

Labour costs can be much less in the Far East than in western Europe and they play a big part in general production costs. However, every other aspect of manufacturing may also be cheaper in the Far East: materials, tooling, plant facilities, and so on.

One expense that may not be cheaper and may, in fact, cost you much more than you bargain for is transportation. If you have to ship your time-critical products by air from China, you're looking at substantial transportation costs. On the other hand, if your invention can take a slow boat from China, that transportation doesn't add a lot. Just make sure that you account for the transportation when you work out your production costs.

Be aware that when you produce your invention overseas, you basically lock yourself into high-volume orders. If a factory in the next town makes your product, you can start with a smaller production run or have an extra 5,000 units made during your next cycle. Such fine-tuning isn't feasible if the production factory is overseas. Making anything less than 10,000 units generally isn't cost-effective, and speeding up delivery is not an option if you base your pricing on shipping costs – switching to air delivery prices you out of the market.

The greatest risk factors in obtaining quality for cost with offshore manufacturing are

- ✔ **Honesty of the contracting brokers:** The cost of raw materials can be many times the manufacturing costs. Should these raw materials become misdirected, the costs to you can be enormous.

- ✔ **Quality control:** The consistency of quality and cleanliness of the finished products can be a problem.

- ✔ **Risk of piracy:** An unscrupulous manufacturer may complete the agreed 1,000 units per day of your product and then make more of the same for the rest of the day, but send these other products along the pirate channels and pocket the sales returns.

If you choose to have your product manufactured offshore, you may want to hire an in-country agent to keep an eye on quality control and piracy for you – but remember, by doing so you add to your unit cost, so be sure to include it in your financial plan.

To find out more about trade links and economic conditions in other countries, visit the UK government's Trade and Investment Web site (www.uktrade invest.gov.uk). It's an indispensable resource featuring a country listing with helpful business guides, as well as customs and regulations information, details about each country's background, and valuable travel advice.

Chapter 18

Preparing for Launch

. . . .

. . . .

*B*eing a developer of ideas resembles being an explorer: You don't know what you're getting into except that that the experience is new, mysterious, and challenging. This chapter shines some light on marketing basics that you need to know in order to prepare your product for the marketplace. We include tips to help you discover whether a market for your product exists and reveal techniques in consumer studies.

Discovering What You Need to Know

Many new inventors assume that everyone is going to love their invention. Unfortunately, that assumption is dead wrong. You may think that because your product is new or because you have a patent, people are going to snatch your invention from the shop shelves. Although a patent gives you an edge, the hard reality is that the big companies have them too. And consumers don't buy something just because it's new unless they need the product.

So, consider finding out what people need and how to give it to them. You have to know who your market is in order to reach it. Basically, you want to answer the following questions:

✔ Who is your customer?

✔ What does your customer need?

✔ How does your invention meet the customer's needs?

✔ Does your invention meet the customer's needs better than your competitor's product?

✔ How can you package your invention to appeal to the customer?

✔ How much money is your customer willing to pay for your product?

Getting the answers to these questions involves gathering facts and opinions in an orderly, objective way to find out what people want to buy. The next sections help you to start the process.

Finding out about your customers

The goal of marketing is to establish customers to buy your product. Ultimately, in order to succeed, you must attract and retain a growing base of satisfied customers. To do that, you must know who your customers are, where they are, and how to get them to buy your product. Finding and utilising the information you gather is what marketing is all about.

Customers are the be-all and end-all of marketing. Without customers, it doesn't matter how revolutionary, how competitively priced, or how attractively packaged your invention is. Without customers, you may as well not even have an invention.

Consumers exhibit different purchasing patterns depending on their age and gender, among other factors. Religion can also influence buying behaviour think of kosher or halal foods. All these different circumstances mean it's likely that not everyone's going to purchase your product. You need to focus on the person who *is* going to buy it. The faster you figure out who the consumer is, the more profit you can make. How do you do this? Through market research.

Keeping up with the competition

When trying to place your product, understand that you're in a highly competitive, volatile environment; therefore, you need to know and understand your competition. And don't say you don't have any competition because your patented invention is unique in the marketplace. You're still competing for customers' money. Consumers have plenty of other opportunities to spend the same pound that you want from them.

You can gain a lot of insight from your competitors – you hope to gain some of their market share, don't you? Even before that, though, you can take a look at their product and analyse its strengths and weaknesses. Use your analysis to make improvements to *your* product before it gets to market. You can also take advantage of the market research that your competitors have done when you decide how to package your invention, what colour to make it, where to sell it, and how to price it.

Start a file on your five nearest direct competitors and five close substitutes. Every quarter, or more frequently if you want, review each file and

- ✔ Evaluate each competitive product's strengths and weaknesses.

- ✔ Estimate whether each competitor's share of the market is growing, staying steady, or declining.

- ✔ Check pricing strategies and determine whether to adjust your current pricing strategy.

- ✔ Look for ways to adapt your competitors' successful strategies for marketing and advertising to benefit your business.

- ✔ Study each competitor's promotional materials and sales strategies.

Visit your competitors' stands at trade shows and look at their presentation from a potential customer's point of view. What does the stand say about the company? Observing which specific trade shows or industry events competitors attend provides information on their marketing strategy and target market.

Many inventors make the mistake of only being interested in their own products, ignoring even the industry in which they sell their products. Doing so is a good way to make sure that the market ignores you all the way to bankruptcy!

Packaging for appeal and profit

Packaging can help sell your invention or it can ensure that your product gathers dust on the shop shelves. Think of 'Pet Rocks', the brainchild of American entrepreneur, Gary Dahl. It was nothing but a packing and marketing genius! Even consumers dense as the proverbial box of rocks knew that there were thousands of stones available for the picking, but the 'inventor' made a mint through packaging. That packaging, which the company designed to look like a pet carrier, enticed people to buy their own Pet Rocks.

Packaging is a key feature in selling your product: It can make or break a sale. Studies show that people look at a product for just a few seconds. So you've just seconds to convince each customer to buy your product. If they can't figure out what your invention is or does in those seconds, they generally put it down and don't think of it again.

You have to find out what type of packaging appeals to consumers and take those qualities into account, along with your product's needs for protection or display.

Pricing your product to sell

Everybody understands that price plays a large part in consumers' purchasing decisions. So spending time figuring out the right balance between the amount customers are willing to pay and the price you need to make a profit is crucial.

Getting the price right

Pricing your product correctly is vital for maximising your revenue. You need to fully understand the market for your product, the channels of distribution, and the competition *before* you establish prices. You must know all the component costs for manufacturing and carefully analyse them. You also need to know the packaging and shipping costs.

Pricing isn't as straightforward as you may think. Many entrepreneurs look at the cost of manufacturing, figure in the cost of packaging, transportation, and overheads, and then try to tell consumers what price they must pay for the product. These entrepreneurs do the process backwards. You don't tell the consumer what to pay for your product; the consumer tells you. In our free market, you can, of course, choose any price for your product. However, the best strategy is pricing your product at what a consumer is willing to pay, and then backtracking all the way down the manufacturing ladder to see whether you can produce your product cost effectively. If not, stay at home. Don't enter the market and lose money.

Selling a product involves two basic costs:

- ✔ **Cost of goods:** The price you pay for manufacturing, packaging, shipping, and handling.

- ✔ **Operating expenses:** These include salaries, advertising, rent, utilities, office supplies, and insurance for both the business and its employees. The more of your product you sell, the more profit you make, as these costs stay relatively constant.

Looking at who gets what

Remember that a product passes through many stages on its way to the shop shelf: A factory shapes the product; the distributor transports it; a sales representative sells it to the retailer; and then advertisers spread the word so the buying public knows that it exists.

Figure 18–1 illustrates where the pounds and pennies go. Our example follows traditional distribution steps and is merely a guideline.

Retail (or list) price	£20.00
(based on research)	
Discount to retailer	
(40% off list)	– 8.00
Retailer's cost	12.00
(From wholesaler)	
Discount to wholesaler	– 4.20
(40% to 35% off retailer's cost)	
Wholesaler's cost	7.80
(From manufacturer)	
Manufacturer's agent	– 0.39
(Based on 5% of Mfg.'s selling price to wholesaler as shown at £7.80)	
Inventor's Royalty	– 0.39
Manufacturer's Net Proceeds from sale	£7.02

Figure 18–1:
Adding up
the costs of
pricing.

Out of the £7.02, the manufacturer must cover the cost of the goods and operating expenses, *and* make a profit! If the manufacturer needs a mark-up of 40 per cent, the product cost (including packaging) must be no more than £5.02. If the manufacturer needs a mark-up of 30 per cent, the product cost must be no more than £5.40.

To get a general idea of mark-up ratios, look at the retail price of a product and divide by 4. For example, if a product sells for £10, the manufacturing, packaging, and shipping costs must be under £2.50. A ratio of retail price four times manufacturing cost is generally the minimum mark-up on a product.

The mark-up on products varies upward from here by industry and product, and also by where the product sells. Get to know the mark-ups for your industry. Manufacturers generally mark up electronic and jewellery products up by a multiplier of 10. For pharmaceuticals, the multiplier may be up to 16. If a product sells on television, the multiplier tends to range from 5 to 7. If the profit margins aren't there for everyone who handles your product to make money along the way, you lose money in the long run.

If a product sells on a home-shopping television channel for £19.95 (and with a mark-up of 6 to 1, meaning a 600 per cent mark-up from the manufacturer to the retailer), the product's manufacturing, packaging, and shipping should be no greater than approximately £3.35.

Conducting Market Research

Doing *market research* means gathering and analysing information about the marketplace or a specific market. Marketing people talk about two basic forms of market research:

- **Primary research** is asking customers or potential customers about their likes and dislikes, as well as other pertinent statistical information. (We explain the process of doing primary research in the following sections.)

- **Secondary research** is making use of existing data. Someone else has already collected (and sometimes analysed) the information; all you have to do is find it and apply it to your situation. The information may be in surveys, books, and magazines.

You can find secondary research material in libraries, universities, trade and general business publications, national and local newspapers, and on the Internet. Good sources of secondary research include:

- **British Library (www.bl.uk):** Based in London, the British Library accommodates an invaluable Business and IP Centre where you can conduct your own extensive market research and gain free access to commercial market research reports.

- **Commercial market research:** Companies such as Mintel, KeyNote, Euromonitor, and Datamonitor publish market reports, which you can buy.

- **Internet search engines:** A great source of information. Try searching for market information using Google, MSN, Yahoo!, or Ask.

- **Local library:** Why not visit the business section of your local public library? You can probably access published market research reports for free.

- **Office for National Statistics (www.statistics.gov.uk):** Economy statistics and information on the UK population.

- **UK Trade & Investment (www.uktradeinvest.gov.uk):** A valuable source of international information if you want to export your product to other countries.

- **UK Trade Association Forum (www.taforum.org):** A national directory of trade associations and business sectors with a useful search facility.

Bear in mind that market research is not an exact science. It deals with people and their constantly changing feelings and behaviours, which countless subjective factors influence.

To do a thorough job of researching customers' likes and dislikes, use the process outlined in the following sections.

Defining the opportunity or problem

People often overlook defining the problem or opportunity, yet this is a crucial first step of the market research process. You have to figure out what your chance is before you can take it. Is there a gap in the market that your invention can fill? Does your invention perform better than any competitor's, yet cost the same?

Finding the root cause of a problem is harder to identify than its obvious symptoms. For example, a decline in sales is a problem, but in order to correct it, you must find the reasons for the decline and correct them.

Setting objectives, budgets, and timetables

Remember that your market research needs to be specifically about your product and cover every aspect of your invention. You want to find out who wants to buy your product; how much they're willing pay for it; what they think that you should call it; how they want you to package it; how they want it to look and function; and where they want to get it. Most important, find out what they don't like and what they want you to change about your product.

Selecting research types, methods, and techniques

Good research reveals pertinent consumer information. You can conduct product research in a number of ways:

- ✔ **Surveys:** Also known as questionnaires, surveys are a time-honoured and effective method of research. You can question people in person, over the telephone, or online.

- ✔ **Field experiments:** Bringing your actual product to people and letting them sample it is a direct way of getting honest reactions. Think supermarket samples and taste tests.

- ✔ **Focus groups:** You get several individuals together, show them a product, and ask them to tell you what they like and don't like about it, what to change about it, and so on. Results from a focus group give you more details than you get from a quick survey; however, it may cost you about the same, if not more, in the long run, because you need to run several different groups.

Designing research instruments

Research instruments is a fancy way of saying *questions*. The following sections help you figure out what questions to ask and how to find people to both ask and answer them.

Asking the right questions in the right way

Poll potential customers by using a survey. Design your questionnaire to find out potential consumers' perceptions of your product.

In developing a survey form for your product, you want to set things up so that the first answer you get is 'Yes'. You design the survey in this way so that it's more user friendly and gets the survey respondents to continue completing the questionnaire. The first question doesn't even have to be about your product. For example, if you're researching a new bicycle accessory product, your first question may be 'Do you like cycling?'.

Then you progress to detailed questions about your product to find out how people perceive it. Perception is everything. You want to know and fulfil the customers' needs. You want to know their colour preferences, packaging preferences, whether the size of the product is correct, what price they'd pay, and where they'd purchase your product.

You already know what you think, but when it comes to market research, what you think doesn't matter. What matters is what potential customers think.

You also want to know who your customer is, so ask a couple questions at the end about age range, income range, post code, buying habits, and so on. Always offer ranges for sensitive information such as age and income level. And always thank your participants.

Figure 18–2 shows a sample survey designed to solicit opinions about a new device (the device in question assists elderly people when taking their medication). Feel free to use this questionnaire, changing the wording to fit your product.

Restrict your questionnaire to three pages, as you have to get the answers you need in a very short time. Give the survey recipients spaces or boxes to tick and a place to add comments if they want to. Limit open-ended questions – you want answers you can code easily. But don't forget one very important open-ended question: Can you think of any ways to improve the product? Pay attention to the answer – write it down and then see how many survey respondents give that same response.

CONSUMER QUESTIONNAIRE
AUTOMATIC PILL DISPENSER

1. Do you use medication on a daily basis?
   _____ Yes   _____ No

2. Would you buy a product that reminded you to take your pills, how many to take, and dispensed your medication? _____ Yes   _____ No
   If no, why not? _____

3. Now you have seen a picture of the "Automatic Pill Dispenser." Would you buy the product?
   _____ Yes   _____ No
   If no, why not? _____
   If yes, would you purchase it for _____ Yourself _____ Gift _____ Other
   (whom) _____

4. How much would you pay for this automatic pill dispenser?
   _____ £ 3.00–£ 3.50   _____ £ 6.01–£ 6.50   _____ £ 9.01–£ 9.50
   _____ £ 3.51–£ 4.00   _____ £ 6.51–£ 7.00   _____ £ 9.51–£ 10.00
   _____ £ 4.01–£ 4.50   _____ £ 7.01–£ 7.50   _____ £ 10.01–£ 10.50
   _____ £ 4.51–£ 5.00   _____ £ 7.51–£ 8.00   _____ £ 10.51–£ 11.00
   _____ £ 5.01–£ 5.50   _____ £ 8.01–£ 8.50   _____ £ 11.01–£ 11.50
   _____ £ 5.51–£ 6.00   _____ £ 8.51–£ 9.00   _____ £ 11.51–£ 12.00
   More than £12.00?_____ If so, what amount? _____

5. Where would you expect to buy this product?
   _____ Sansbury's _____ Tesco _____ Asda _____ Boots
   _____ Television _____ Superdrug _____ Supermarket pharmacy
   _____ Independent pharmacy _____ Mail order
   _____ Mail Order Catalogue _____ Magazine Ad _____ Newspaper Ad _____
   _____ Other (please name) _____

6. Would you prefer to purchase in specific colours? _____Yes _____No
   If yes, what colours? _____

7. How would you expect to see this product packaged? _____ Box _____ Hang tab
   _____ Bubble pack _____ Other, please name _____

8. After seeing the "Automatic Pill Dispenser", can you think of any ways to improve the product? _____ Yes _____ No If yes, what changes would you make?

9. If this were your product, what would you name this product? _____

10. Personal information:
    Age: _____<20 _____ 21–25 _____ 26–29 _____ 30–34 _____ 35–39
         _____ 40–44 _____ 45–49 _____ 50–54 _____ 55–59 _____ 60–64 _____65+
    Sex: _____ Male _____ Female
    Income Level: _____< £15,000 _____£15,001 - £19,999 _____£20,000 - £29,999
                  _____£30,000 - £39,999 _____£40,000 - £49,999 _____£50,000+Additional
    Comments:_____
    _____

Figure 18–2:
A sample product survey.

One of the most important items a survey reveals is what potential consumers don't like about your product. And, that's a good thing. It's just as important to find out what people *don't* like as what they *do* like. Knowing this saves you money in the long run: You can make changes before you launch your product, so you're ahead of the game already. For example, if enough consumers tell you that they want your product in red rather than black, you make it in red. If you find out customers prefer a box to shrink wrap, you account for the cost of a box in your profit margins and sell your invention in a box. Offering a product that consumers want is much cheaper than having to spend time convincing consumers that they like and want what you have.

The results of your market research may surprise you. You may think that a supermarket is the best place to sell your invention, but your market research may show that 85 per cent of the respondents would buy the product via a television shopping channel. Changing the venue dictates changes in the mark-up, packaging, and profit margins.

They say that a picture is worth a thousand words, so make visual arts work for you. It's amazing what a good graphic artist can do with coloured sketches or computer drawings of your product. In many cases, you can get artistic renderings of your product for a small portion of the cost of a prototype. Show the visuals while people answer survey questions.

Getting the right person to ask the right questions

If you're on a tight budget, one of the cheapest ways to conduct primary market research is to hire students from a nearby college or university. People respond more warmly to students and are generally happy to speak to them, whereas professional marketers often get the cold shoulder. If you don't use students, you can call your local Job Centre, use a temping agency, or hire a market research firm.

If you decide to use a professional market research company check it out first. Believe nothing unless you take the time to investigate the company and its past experience in research projects. Get the names and phone numbers of references. As a starry eyed inventor who thinks that your invention is the greatest thing since sliced bread, you're a prime target for getting ripped off by unscrupulous companies that tell you that your invention is, in fact, the best thing since sliced bread, then charge an arm and a leg to do market research that isn't worthy of the name. Take off your rose-tinted specs when seeking market research.

If you conduct your own research, don't identify yourself as the inventor. You can't afford to have people be nice to you and tell you what they think that you want to hear.

Basically, you want any potential customer to answer your survey, except, and it's a very big *except*, your friends and family. You don't want to be one of those inventors who believes that the deal is sealed because 25 of their closest friends and family members told them that the product can sell. Ask friends and family who say your invention is going places to invest in getting the invention to market. Give them the opportunity and see what happens. Most don't invest. If you really want to succeed, you want honesty, not kindness.

In addition to potential customers, survey managers of shops where you think your product may sell. Finding out this information can be an eye-opening and very valuable experience. Shop managers generally rack up several years of retail experience before they become managers. Be aware that when dealing with the large retailers, managers manage people; they've little say about what products the shop carries, because national buyers order for all the shops in the chain. That fact doesn't mean that these managers can't be helpful to you, though. Just like managers of smaller shops who order their own goods, these managers are aware of products in general, what it takes to sell a product, pricing schemes that help sell goods, the importance of packaging, and other hands-on issues that you can benefit from.

Figure 18–3 is a sample manager survey, which you can adapt to your needs. Note that the shop manager survey asks very similar questions to the consumer survey.

Surveying shop managers is a great resource. For example, if you survey shop managers on price and take the average, generally this figure is within a 10 per cent deviation of the actual selling price. For example, if you survey 15 managers and they state your product would retail for £4.95, the error rate would place the retail price 10 per cent higher or lower than that amount. Your product would most likely retail for £4.50 to about £5.50 with the average of £4.95.

Try to get a broad sample of managers. For example, if you have a lawn and garden product, talk to managers at garden centres, large home improvement stores, and local hardware shops that sell gardening supplies.

You can conduct a manager survey yourself. In your survey, you don't ask about your product directly, but about what makes products like yours sell. By all means, bring a prototype or drawings, and cheerfully accept any comments or criticism. But your main role is to get the manager to talk about similar products and how they sell. If the manager talks about flaws in products already on the market, you can probe to discover how you can improve your product.

SHOP MANAGER SURVEY
THE AUTOMATIC PILL DISPENSER

1. What is the policy and procedure for placing products in your shop?
   _____

2. Who purchases the products for your shop?_____
3. What is the address of your head office? _____
   _____
   _____

4. What kind of packaging would you use to display this product?
   Bubble pack _____ Hang tab _____ Box _____ Other _____

5. Do you like this product? Yes _____ No _____ If no, why not? _____
   _____

6. If yes, would you consider selling the product in your shop?
   Yes _____ No_____ If no, why not? _____

7. How much would you charge for this automatic pill dispenser if it were sold in your shop?
   _____ £3.00–£3.50   _____ £6.01–£6.50   _____ £9.01–£9.50
   _____ £3.51–£4.00   _____ £6.51–£7.00   _____ £9.51–£10.00
   _____ £4.01–£4.50   _____ £7.01–£7.50   _____ £10.01–£10.50
   _____ £4.51–£5.00   _____ £7.51–£8.00   _____ £10.51–£11.00
   _____ £5.01–£5.50   _____ £8.01–£8.50   _____ £11.01–£11.50
   _____ £5.51–£6.00   _____ £8.51–£9.00   _____ £11.51–£12.00
   More than £12.00?_____ If so, what amount? _____

8. Who would you talk to within your own company to market a product if you had a similar idea? _____

9. What would you name this product? _____
10. If you were in my position, what would you do next to market this product?
    _____
    _____

11. Do you have any additional suggestions that may help me get this product on the market? _____

12. Do you have any additional comments? _____
    _____

Figure 18–3:
A sample
manager
survey.

Call a shop manager to try to arrange an appointment, but *don't* identify yourself as an inventor. The mental image most people get when they hear 'inventor' is of someone who looks like Albert Einstein and acts like Sherman Klump in *The Nutty Professor* – someone smart but weird who's going to waste their time. Try 'entrepreneur' instead. Say something like, 'I'm a start-up entrepreneur with a new product. I am *not* trying to sell you anything; I just want your expert advice. Would you take a few minutes to speak to me?'

Asking for help is perfectly acceptable, and many people are very willing to give it. Smart retailers know that entrepreneurs form the backbone of the country, and they're often more than happy to make a contribution to the next great thing.

Organising and Analysing the Data

When it comes time to analyse the data gathered from your market research, you quickly realise the value of having tick-boxes or spaces on your question-naire. What you want is a tabulation of how many people chose each response to every question and the total percentage of people who chose each answer. For example, if you ask which colour people prefer and give respondents choices of red, yellow, green, and blue, you want to know how many people prefer each colour and what percentages those numbers translate to.

In reality, generating a report of your survey responses is probably the simplest part of the whole process. You just plug in the responses and out come the numbers.

Use the data you gather to build a *customer profile* that tells you who your typical customer is – age, gender, income, location, shopping patterns – everything and anything that can help you sell your product.

Part V
Developing Your Market

'Unlimited and lucrative royalties for your thoughts, dear.'

In this part . . .

Getting a toehold in the market is just a start to getting worthwhile returns. Your product needs to grow from that start into a viable business proposition. You need to promote your invention, advertise it, and maybe license or franchise it. These chapters show you how to expand the business side of your brainchild, and how to negotiate the licensing process from contact to contract.

Chapter 19

Marketing Your Product

· ·

In This Chapter

▶ Creating a marketing strategy

▶ Keeping yourself on track

· ·

*L*aunching your invention is a foray into the dog-eat-dog world of the marketplace. Your product competes for the consumer's pound against similar products (and purchases in general) to find its place in the world. You can now measure all your work in sales revenues, as you test your invention in the court of last resort – the marketplace. Here, you ultimately win or lose the battle: A purchase is a win; staying on the shelf is a loss. No matter how well developed and one-of-a-kind your product is, if it doesn't sell, it isn't a success.

As an entrepreneur, you face rejection from buyers, wholesalers, consumers – everybody, it seems. Most successful entrepreneurs go through the hard times before enjoying success.

For every pound you spend on development, spend ten pounds on marketing. A good product properly marketed wins over a great product poorly marketed.

Everything you do in development, design, protection, branding, positioning, and pricing culminates with marketing your invention and selling it to your customers. This chapter reveals what's involved in marketing your product to potential customers.

Developing Your Marketing Strategy

You don't build a house without a plan, do you? Of course not. Likewise, you need a plan to sell your invention. A sound marketing plan is key to the success of your product.

Marketing is civilised warfare, so get ready for the fight of your invention's life. You fight for the same shelf space, magazine space, and TV time as firms with multimillion-pound advertising budgets. It's not so much that the shelf ain't big enough for the two of you, but that products already on the shelf squeeze into a smaller space to make room for yours. If that happens with a national chain, break out the bubbly because you've reason to celebrate. On the other hand, getting your product into major stores is rare enough that you don't have to feel a failure if you don't get your invention placed straight away.

A *marketing strategy* identifies customer groups (known as *market segments*) that your business can serve better than your competitors can, and tailors prices, distribution, promotions, and services towards those groups.

Your marketing strategy accomplishes two essential tasks:

- ✔ **Directing your company's policies and activities toward satisfying your customers' needs:** Every customer has specific needs. Marketing determines these needs, and you develop the product to satisfy them. Market research, focus groups, surveys, and questionnaires are crucial to determining customers' needs.

  The most successful products fill *needs* rather than wants or desires. A need is something you absolutely have to purchase; a want is something you'd like to have. For example, if you don't have access to public transport, you *need* a vehicle. You may *want* that vehicle to be an Aston Martin, but then that's the difference between needs and wants.

- ✔ **Pricing your product for optimum profit:** Realise that making a profit is more important than maximising sales volume. You may sell millions of units, but if you don't make money, you waste your time and effort.

Your marketing plan includes your market research (see Chapter 18), your location, the customer group you target, your competition and positioning, the product you sell, pricing, advertising, and promotion.

From a thousand different approaches and even more methods, marketing programmes all aim to convince people to try out or keep using particular products or services. Analyse your product's competitive advantages to develop long- and short-term marketing strategies. Also, carefully plan your marketing strategies and performance to establish a market presence and keep it strong. You need to establish brand loyalty so that customers come back time after time, bring their friends and family to you, and eagerly purchase any new products that you develop.

Staying on top of customers' needs

Use these tips to keep abreast of customers' needs:

✔ Complete at least one marketing activity each day, such as conducting a follow-up call with a customer, e-mailing a product rep, or writing a thank-you note to a supplier.

✔ Establish a way of tracking your clients and keep your list updated.

✔ Calculate a percentage of your company's income to spend on marketing functions each year.

✔ Set your marketing goals for the year but review them on a quarterly basis to make sure that you meet them.

✔ Keep business cards with you at all times – you'll definitely run into numerous occasions to distribute them.

Mixing it up

Every marketing strategy combines four key components, called the *marketing mix*:

✔ **Products and services:** In your case, your invention.

✔ **Promotion:** Basically, any method you use to get your product into the public's eye and consciousness. Promotion includes advertising, customer relations, leaflets, and so on. Read more about advertising in Chapter 20.

✔ **Distribution:** The method you use to get your product to your customers. A mail-order business, not surprisingly, distributes through the mail. A retail business, however, may use a combination or a chain of distribution methods – factory to lorry to ship to rail to lorry to distribution centre to lorry to shop to customer, for example.

Distribution includes storage as well. If you have to pay to store your goods, you have to account for those costs.

✔ **Price:** Getting the price right is crucial. You have to determine how much customers are willing to pay and figure out how to make a profit for yourself based on that price. See Chapter 18 for more on pricing.

The following marketing tips can help you stay sane in the marketing madhouse:

✔ **Use your customers.** Ask them how they found out about you and why they decided to purchase your product. Ask them for marketing ideas – maybe even run a contest. Definitely ask any customers who return your product their reasons for doing so.

✔ **Keep up with the industry.** Read trade magazines, subscribe to trade journals, and join the trade association. Subscribe to any Internet newsgroups that deal with your industry. Attend educational events and conferences. Join professional organisations.

✔ **Keep up with your competition.** Keep and update files on your competitors' ads and marketing materials. You can discover a lot from looking at material about their product features, marketing strategy, pricing information, and packaging material.

✔ **Hire a marketing consultant to generate new marketing ideas and suggestions.** A marketing expert most likely sees your product in a different light and has links, resources, strategies, and comments you may not think of. The services of an experienced marketing consultant may be expensive, so ask for references before signing one up.

Targeting your market

If you're like most inventors and start-up entrepreneurs, you've a limited market budget. Concentrating your efforts on one or a few key market segments – *target marketing* – can save you money and help your business flourish.

In order to target a specific market, you need to identify it and then break it down or segment it. You can do so in a couple of ways:

✔ **Location:** Specialise in serving the needs of customers in a particular geographical area. For example, a restaurant may send advertisements only to people living within a 10-mile radius.

✔ **Customer type:** Identify those people most likely to buy your product or service and target them. Remember, not everyone wants to buy your product. For example, if your product is a new baby monitor, target parents-to-be, new parents, and grandparents, and don't concern yourself with people who don't fit into those segments.

Narrowing your market helps you save costs on advertising and serve your customers better. The more precisely you can define your market, the more specifically you can gear your product to meet your customers' exact needs.

Making sales presentations

One of the most effective ways to market your invention is with the personal touch – making sales presentations. If you're good at selling, that's great. If you're not much of a salesperson, practise until you are or get someone who's good to present for you.

You (or someone else, if you can't bear the thought of selling) can sell to the customer in one of two ways:

- ✔ **Direct selling:** You sell your product directly to the customer through a face-to-face transaction, or via mail or the Internet. You take home the entire selling price and deduct your costs from that to determine your profit margin. Direct-sales venues include craft fairs and festivals, car-boot sales, your own shop, mail-shots, or your company Web site.

- ✔ **Indirect selling:** You pay someone else to sell your product. You pass the cost of the salesperson on to the customer. Indirect sales methods include hiring sales representatives (who sell a number of different products and receive a commission on each product sold), who place your product in a shop to sell on consignment or sell your product direct to a shop. A typical mark-up is between 60 and 100 per cent, which is what the shop needs in order to cover its operating costs and contribute to its profits.

You can sell directly to shops, but you may have more success getting an appointment with a sales representative or *vendor* (supplier or seller of goods). Source the contact details for a key retail buyer and call them. Don't be offended if you don't get a return phone call. You may have to make several more before a buyer calls you back – they're very busy people. When you finally reach a buyer, be friendly. Don't put him on the spot by asking for a commitment, or even an appointment, then and there. Instead, ask him for the names of the sales representatives that he deals with most frequently. The sales reps are the people you really want to sell on your invention.

One of the easiest ways to get your product onto the shelves is to go to a shopping centre, department store, or supermarket and search for similar products. Sales representatives already sell these products in the shops. If a product is already for sale in the shop, it already has a *pre-approved vendor*. Find out who the sales rep and distributors are and then call them and see if they're willing to take on your product. Sales reps sell on commission and they want to sell new and exciting products.

Different companies have different sales structures, so make sure that you don't bypass a buyer in making a sales pitch to a board of directors. Going through the right channels makes you more popular and makes it more likely that the company listens to you.

Sales presentations may feel like the last thing you want to do. But what's the worst that can happen? You may get thrown out, you may not sell your product, or you may be insulted. So what? Just remember: The more presentations you make, the higher your probability of getting sales.

When you do make a sales presentation, dress accordingly in business attire.

Attending trade shows

One of the most important things you can do to market your product is to attend trade shows and exhibitions. A *trade show* is an event held in an exhibition centre such as the NEC in Birmingham or Earls Court in London, where manufacturers and marketers display their products and technologies to potential buyers, retailers, wholesalers, and others in their industry. Trade shows provide meeting grounds for an interchange of new product information between buyers and sellers.

Tracking down trade shows

Nearly every industry has a trade show. Sometimes, the show brings in famous speakers and celebrities in order to attract a bigger crowd. Also, the show may offer educational events, with a variety of speakers and topics targeted toward a specific industry.

Trade shows generally take place at a single location – often at the same location each year. They usually last one to three days, and bring together thousands of exhibitors and potential customers. These shows give you a powerful way of selling your product, as well as finding out more about the industry.

Trade shows offer an inexpensive way to meet many potential customers face to face in a brief period of time. But be aware that your potential customers at a trade show are retail chains, not individual consumers. You can use the same marketing techniques; you just sell on a larger scale.

The inventor who asks, 'Where do I find out about trade shows, how much do they cost, how do I get in, what do I do when I'm there?' has a higher probability of success than an inventor who wants nothing to do with the hustle and bustle of a trade show. In fact, make going to trade shows part of your annual business plan (Chapter 14 covers business plans).

At www.exhibitions.co.uk you find a useful Web site that lists the names, addresses, activities, and publications of exhibitions and their organisers. Use the search function to help find trade shows relevant to your business.

Deciding whether to exhibit

If your budget is down to the bare minimum, regard trade show fees as a necessity *not* to be cut. Trade shows help level the playing field for smaller firms. Stand space is generally inexpensive and is sold in square metres, so you can choose the right space for your needs and budget. With creative marketing and stand design, small businesses can actually appear as substantial as much larger organisations.

If you just want to attend a trade show, sometimes the only cost is for stationery (taking a business card or letterhead with you) that proves that you're in the industry. Other shows are free if you pre-register online. Otherwise, the entrance fees are normally less than £25.

Both exhibiting and attending have their advantages, which we summarise in Table 19–1.

Table 19–1	Exhibiting Versus Attending
Reasons to Exhibit	*Reasons to Just Attend*
Generate sales leads and actual sales.	Meet buyers, wholesalers, distributors, and potential licensees for your product.
Enhance your image and visibility.	Find out how the industry works.
Reach a specific audience.	Check out the big names' swanky stands and see what you're up against when you decide to exhibit.
Establish a presence in the marketplace.	Become familiar with the language of buyers and sellers. Even the best sales reps practise the lingo in order to become more proficient.
Improve the effectiveness and efficiency of your marketing efforts.	Circulate and see the sights instead of being stuck on one stand for the whole show.
Personally meet your customers, competitors, and suppliers.	
Prospect for new customers.	
Introduce a new product.	
Demonstrate your product.	

If you're an independent inventor, you may not want to spend the money for a trade show stand. Just be there. Anyway, if you're on your stand and by yourself, you can't leave it – not even to grab a coffee. If you're on a limited budget, you may end up with an 8-foot table, covered with a white tablecloth and a selection of home-made posters. That doesn't sound too bad until you realise you may spend three days right next to a stand with £500,000 displays!

Planning and preparing to make the most of your time there

Sophisticated, well-prepared exhibitors do well at trade shows no matter what the size of their business, and the naive and inexperienced can waste thousands of pounds and countless hours – and maybe do more harm than good.

However, using trade shows effectively takes only a little effort and planning:

  ✔ **Come equipped, mentally and physically.** Mentally, be prepared to put your best foot forward at all times. Be friendly, interested, and responsive. Physically, make sure that you wear comfortable shoes (plan on a lot of walking on convention centre concrete floors). Come prepared with business cards to give to everyone you meet, and bring a bag to store all the literature and free handouts you pick up. If you take a sample of your product to the exhibition for presentation, consider investing in a professional case with a handle and wheels to create the right impression and take the strain off your arms!

  ✔ **Plan a course of action.** Make note of any events you want to attend, and locate any exhibitors whose booths you want to see. Target the people you want to meet and make appointments ahead of time. You can meet buyers, wholesalers, distributors, manufacturers, and potential licensees for your product.

  ✔ **Take advantage of the unique opportunities a trade show offers.** You can compare dozens of competing vendors, handle new products and supplies, and test products too large or cumbersome to be brought to the office. You also get to see new inventions (of course) long before they hit the market. Trade shows really provide a showcase for brand-new products, technologies, and trends in an industry. Many inventors hold back a new product so that they can launch it at a trade show.

Making contacts

A trade show is a giant networking opportunity. You can speak to competitors, clients, suppliers, and manufacturers in the space of a couple of days. Take advantage of the opportunity to speak with technicians and designers who are directly involved with your competition. Don't be shy about introducing yourself and offering to buy a beer or cup of coffee for someone whose brain you want to pick. The cost of that drink can be a great investment.

You may want your product to sell in one or more of the catalogues, trade journals, or newsletters you sign up for. Bring business cards so that you can get on mailing lists, not only of wholesalers, buyers, distributors, and suppliers, but also your competitors. You can keep ahead of your competition – see what they do and what new products they introduce.

Follow up those contacts you make, ideally as soon as possible while their details stay fresh in your mind. Spend time after you get back sending e-mails, making phone calls, and sending any information or samples you promised.

Gathering ideas

Take advantage of the vast amount of knowledge and experience on offer to expand your own understanding of the industry. Your colleagues are fellow trade show participants and their opinions and experiences can help you make decisions about your product and future shows you want to attend.

Find out who the major players are, what they're working on, and where they think the industry is heading. You can focus on understanding the buyer–seller relationship because wheeling and dealing is going on all around you. You can see what sells now and get ideas about what may sell next year. Keep your eyes and mind open.

Generating sales leads

Because business-to-business shows typically don't allow selling on the show floor, generating sales leads is the most common reason exhibitors participate. During the course of one trade show you may personally meet most of your important clients and suppliers, making shows a good way to establish and reinforce relationships.

At trade shows, you get contacts and sales leads and possibly take orders; however, no one holds a big inventory in stock – you don't see articulated lorries unloading merchandise. Depending on the venue, you may be able to take orders, but the show probably won't allow you to have the products for consumers to take home with them that day.

Trade shows generally only open to players in the industry and not to the general public. Wholesalers, distributors, and retailers give out information and price sheets, but don't do much actual selling. Industry buyers who attend often represent many shops in the chain, not just one or two, and it's not cost-effective for them to take small orders from start-up companies who may not be able to deliver what they promised.

Examining the extras

Other trade show extras include special events, promotions, and incentives that encourage attendance and sales on the floor. Free training seminars and workshops provide great opportunities to network with colleagues who have similar interests. Vendor-sponsored parties provide an enjoyable social outlet, free from selling pressures and full of contacts. Trade show attendees also have the opportunity to enter a multitude of draws for a variety of interesting prizes.

Bringing your product and customers together

Getting your product into your customers' hands involves the two Ds: distribution and display, both of which we cover in the following sections.

Distributing widely and well

Distribution is the manner in which you get your product to the customer. You have many factors to consider when it comes to distribution. You need to have a well-thought-out plan, and answering the following questions can help you devise one:

- **How many products can you store?** What's your inventory capacity? Use sales forecasts to decide what your inventory levels should be in order to meet customer demands.

- **What's the turnover rate for your inventory?** How does it compare with the standards in your industry?

- **Do you have cyclical fluctuations or seasonal changes that affect the demand for your product?** For example, if you produce stand-alone air-conditioning units, how can you manage peak production and sales periods as well as slow periods?

- **What's your distribution channel?** Describe your distribution channel in detail. Plan on a step-by-step basis how your products is to reach your customers.

Every method of distribution has pros and cons. Consider all the advantages and disadvantages of each method of distribution in relation to your situation. Include in your analysis a look at shops, craft fairs, selling from home, mail order, sales reps, trade shows, and wholesale. You may want to use one method exclusively or combine several.

Make sure that you figure out the cost for each distribution method you consider and add that into your pricing calculations.

For an item for which you anticipate a relatively small sales volume, the distribution method is such a routine part of the sales contact that it requires only a small amount of thought and description. In this situation, you display your baskets so that consumers can see and purchase them. In other cases, you may want to sell your baskets over the Internet, where you ship your product directly to the end purchaser. The point is that you just want to get your product in your customer's hands and receive payment.

Your method of distribution depends on you and your product. Start with a single system so you can get experience in distributing your product. As your knowledge base grows, you can add additional products to product lines, as well as broaden your customer base.

Whatever you do, deliver your product when you say you will. Keep your word. You don't want to promise what you can't deliver. Negative word-of-mouth advertising spreads fast and does more damage than positive advertising can fix. Regaining credibility is very difficult.

Displaying your product to advantage

Getting customers to buy your product often boils down to how it looks in the shop. If your packaged invention takes a large amount of display space, shops may be reluctant to stock it. Remember, shelf space is expensive. If your invention doesn't stack, you place it at a disadvantage from the outset. A large or heavy item requires special handling. You have to keep all this in mind to give your invention its best shot.

Placement is important too. You don't see name brand products stocked on the bottom shelf of a shop. Big companies pay to have their merchandise stocked at eye level. They also pay to sell their product at the end of an aisle. If your product is an impulse-buy item, such as chewing gum or breath mints, you want the shop to place it in a high-traffic, high-visibility area, preferably right by the tills. Location is less of a concern for products that customers are willing to go out of their way to find. If someone needs cold medicine or a special type of tool, that person looks for it and finds it.

Steering clear of invention promotion companies

Don't end up being one of the countless inventors who pay thousands of pounds to fraudulent invention promotion companies. Fraudulent schemes have deprived national economies of thousands of new products and technologies for decades. Unsuspecting inventors paid millions of pounds to these con artists and got little or nothing in return.

Fraudulent invention promotion companies just want to get their hands on your money. They don't want to get your invention into production and they won't help you become a famous inventor. They just want your money: plain and simple.

Widening the gulf

It used to be that manufacturing firms were eager to talk to inventors. That's not quite the case any more. What led to the breakdown? Well, a significant factor is the growth of fraudulent invention promotion schemes. These opportunists flood the system with so many unrefined and undeveloped ideas that many manufacturers don't even open their mail any more.

However, just closing down the fraudulent schemes won't mend the system alone. Other forces are at work. You, the independent inventor, also contribute significantly to this breakdown.

An underlying problem here is that many independent inventors have ugly babies (inventions). No one likes to hear their baby called ugly, and people don't want to hurt your feelings if they can help it, so far too few inventors get an honest appraisal of their invention. On the other hand, though, the overwhelming majority of inventors just wish that someone had been honest with them very early in the process – before they invested their life savings – and told them that their baby was indeed ugly.

To deal with the problem, inventors need to establish a mechanism very early in the process that can screen out the truly unworkable or unmarketable inventions. The mechanism must be affordable, unbiased, and 100 per cent ethical. Refer to Chapter 12 to find out how to evaluate your own invention's potential or to employ the services of a legitimate company to conduct a professional assessment on your behalf.

Watch out for fraudulent invention marketing company ads; they're everywhere – in the back of magazines, in newspapers, on television and radio commercials, and any other media naive enough to take these companies' money.

Don't use the services of or listen to the advice of an invention promotion company without first carefully checking the company and its successes. Our general advice is to steer clear of them entirely. Continue to educate yourself and continue to develop your idea; just don't let your enthusiasm for that idea overpower your common sense.

Evaluating Your Ongoing Progress

After implementing a marketing strategy, you must evaluate your product's performance. Every marketing programme should have performance standards to compare with actual accountable results. Researching industry norms and past performance helps to develop appropriate standards.

Monitor population shifts, legal developments, and the local economic situation to quickly identify problems and opportunities. Also, audit your company's performance on a quarterly basis. The key questions to ask yourself are:

✔ **Is my company doing all it can to be customer orientated?** Think of a young person who mows lawns to earn some extra pocket money. After a few months of mowing lawns, he gets a friend to call his regular customers and ask them if they're happy with their current mowing service. The customers think that they're speaking with someone from a different mowing service, so they're usually honest. What a great way to see whether your customers are satisfied with the service you provide.

✔ **When customers purchase a product, do they want to come back or at least recommend my product to others?** Your current invention may be the first of a whole range of new products. How do you go about making customers loyal to your brand?

✔ **Is my product selling at a competitive price?** If it isn't, you can guarantee that your competitors' products are. One of the best ways to keep wholesalers, distributors, and retailers selling your product is to be the most competitive that you can be so that others won't even think about entering into the market.

✔ **Who are my prospective customers?** Always look to expand your target customer base. You can use ongoing market research (see Chapter 18) and current market trends to identify potential customers.

You may be doing market research without being aware of it. You look at returned items, ask former customers why they switched products or brands, and check your competitors' prices, don't you? Your ongoing research helps you identify areas for improvement and expansion.

Chapter 20

Advertising Your Product

· ·

· ·

*T*his story has been told in different versions through the years:

> A man wakes up in the morning after sleeping on an advertised bed, wearing his advertised pyjamas. He bathes in an advertised bath; washes with advertised soap; shaves with an advertised razor; eats a breakfast of advertised fruit juice, cereal, and bread (toasted in an advertised toaster); and glances at his advertised watch. He drives to work in an advertised car, sits at an advertised desk, uses an advertised computer, and writes with an advertised pen.
>
> Yet, the man hesitates to advertise, saying that advertising doesn't pay. Finally, when his business goes under, he advertises it for sale.

The point, in case you missed it, is that nearly everything is advertised. And advertising is everywhere. You, and every other consumer, respond to advertising whether consciously or unconsciously. You, and every other entrepreneur, need to use this powerful tool. This chapter tells you how.

Exploring Advertising Basics

Advertising serves a very basic function – it tells potential customers about your invention. People have to know that your marvellous product exists in order to buy it.

Advertising is how you communicate with consumers, companies, and potential purchasers. The buzz term is *marketing communications*. These communications help you to keep in touch with your market. You also get a chance to present your product, your image, and your message in the way you want. You've complete control over what's published, broadcast, and displayed, and you determine exactly where, when, and how often your product's message appears, how it looks, and precisely what it says.

Through your adverts, you establish a unique identity for your company and product. You set yourself apart from your competition, and you also remind current customers and tell potential customers about the benefits of your product. And you keep on telling them. In order to keep your customers from spending their money elsewhere, you must remind potential customers repeatedly about the benefits of buying your product and doing business with you. If you don't, another company does.

Generally, a consumer sees a product several times before purchasing it. So design your ad campaign to enhance your reputation and cement your image in the public's mind. To do so, you need to present a consistent image and message repeatedly. People like what they know. The more you can keep your product in the shoppers' minds, the more comfortable they become with the product, and the more they want to buy it.

Although advertising can do many good things for your invention and your business, it's *not* a cure-all. For instance, advertising can't

- Create an instant customer base
- Cause an immediate sharp increase in sales
- Solve cash flow or profit problems
- Substitute for poor or indifferent customer service
- Sell useless or unwanted products or services

You have to start with a solid product and a sound business plan before your advertising can truly be successful.

Designing Your Advertising Campaign

Your advertising campaign must be fluid and adaptable to changes in your product and in the market, but it must also be cohesive and send a consistent message. The specifics can (and should) change, but the basic elements of an effective advertising campaign are:

> ✔ Theme
>
> ✔ Audience
>
> ✔ Range of media
>
> ✔ Objective with goals that you can measure
>
> ✔ Budget

Say, for example, that you launch a brand-new product. Your theme incorporates the product's benefits, but the main theme is that the product is new. You may start out with a target audience that focuses on existing customers, but remember that your campaign may expand to include a new target audience. The range of media includes radio spots, a series of newspaper ads, and a direct mailing to your customer list. Your goal is to sell 100 units by the end of your two-week media blitz. Finally, don't forget that you're on a budget. Work out how much you can spend and stick to it.

The following section explores in more detail each element of an advertising campaign.

Tying into a theme

The first step in creating your advertising campaign is to establish a theme that identifies your product in all your advertising materials. The theme of your advertising reflects your business or product's special identity or personality and its particular benefits. For example, cosmetics ads almost always rely on a glamorous and sexy theme. Automotive advertising frequently concentrates on how the car makes you feel when you drive it.

Cutting into the competition

One family barber shop took a very creative advertising route. This old-fashioned shop, complete with red and white pole, was losing its customers to a trendy new hair salon in the adjoining shopping centre.

The new styling salon was part of a national franchise, with a significant advertising budget. The salon advertised its £25 haircuts on billboard after billboard leading up to the shopping centre. The established barber shop had an almost non-existent advertising budget. However, it did drum up enough money for one billboard advertisement. Placed next to the competition's final billboard outside the shopping centre, it said simply: 'We FIX £25 haircuts!' Short, simple, and effective!

Keep your message simple so that customers can quickly and easily understand it. Consumers want to know that your product works, not necessarily *how* it works.

A distinctive image, whether it's a logo, trade mark, or catchphrase that customers clearly associate with your company, is an invaluable selling tool. Customers recognise your product quickly and easily – in ads, mailings, packaging, or signs – if you use a consistent and distinctive message.

A catchphrase or slogan reinforces the single most important reason for buying your product. For example, 'Nothing Runs Like a Deere' (John Deere farm vehicles) implies performance and endurance with a play twist on the word 'deer.' 'It's the real thing.' (Coca Cola soft drinks) again signifies taste, but also reliability and imagination.

Targeting your audience

Through advertising, you can encourage existing customers to buy more of what you sell and attract new customers. Your research into the market provides you with knowledge about who your target customer is (see Chapter 18 for the low-down on researching). To form your advertising plan, you need to figure out how you can best reach those people.

To reach the broadest spectrum of potential customers, use a broad range of advertising methods. Don't put all your money into newspaper ads, radio spots, or any one medium. Studies show that at least three full exposures are necessary to attract a customer to buy a product, which explains why you often see and hear the same advertising campaign for a product on TV, in magazines and newspapers, and on the radio all at the same time. This tactic, known as a *marketing blitz*, is very effective at moving a product into the public consciousness. You may not be able to afford a full marketing blitz, but you can make your advertising budget generate better returns by spreading it around.

You can target particular areas or people, but you still want to use a variety of ways to reach them. Here are a few hints on how to do that:

✔ Send out press releases by e-mail to major newspapers, radio, and television contacts. Find out who the right journalists are, approach them directly, and follow the press release up with a telephone call. See the later section 'Working up some free publicity' for details on how to approach your release.

✔ Write your own stories for the press – make their life easier. People love to read success stories.

✔ Consider writing a letter to the editor of your product's trade magazine.

Choosing the best media for your message

Your choice of advertising media is varied. Aside from the obvious television, newspapers, and radio, a whole host of printed media exists, including magazines, newspapers, and journals. You can also send out mail-shots.

Consider advertising on the Internet. You can pay for ads on search engines and other people's Web sites, or design your own site to advertise and sell your product.

Your target audience helps determine the media you use to reach it. When you conduct your initial market research, consumers tell you where they anticipate buying your product from. For more on market research, see Chapter 18.

Table 20-1 gives you some suggestions for where to place different types of ads, which, for lack of a better system, we classify as small, medium, and large. Small ads are business card size, and sometimes that's exactly what they are – a reproduction of your business card. Medium ads are the size of a piece of A4 paper, and large ads are poster size. Note that ads in newspapers and magazines can be any size, depending on your objective and budget.

Table 20-1	Suggestions for Ad Placement	
Small	*Medium*	*Large*
Magazine back pages	Direct mail, or with your invoices	Road-side billboards
Neighbourhood association newsletters	Newspaper insert	On buses
Civic and social organisation newsletters	Leaflets door-to-door and in the street	On recycling containers
Publications sent to graduates from their old universities	Shopping centre message boards	On buildings

Don't forget the *Yellow Pages*, local business publications, and magazines specific to your geographic or product area.

Advertising in the local free newspaper is much cheaper than in the daily national press. Don't forget to send smaller publications your press releases. They often want filler info.

Tearing out Goliath

A small pizza restaurant in America found a way to use its competitor's big advertising budget. The small business had a very small advertising budget. Its major competitor was a national pizza chain with franchises throughout the city. The chain spent lots of money to take out full multicoloured ads in the phone book, listing the phone numbers and locations of all its pizza places.

The smaller business couldn't afford phone book ads or even local newspaper ads on a continual basis. But the small restaurant could afford radio ads that aired around the time that people got their new phone books. The ad promised a free pizza in exchange for the competitor's full-page ad in the phone book.

People tore the ads out of phone books at work, at home, in telephone boxes, and hotel rooms. It cost the small pizza place numerous pizzas during the first few weeks, but for the rest of the year, those pages with the competitor's phone numbers no longer appeared in the local phone books. That's one creative gimmick.

Setting objectives and measuring results

You have to set an objective for your advertising campaign. If you don't, how can you know when you succeed? Objectives make it easier to design an effective marketing campaign and help to keep that campaign on track. Defining your objectives also makes it easier to choose how you want your advertising to reach the customer.

Your objective may be as simple as communicating your message or creating an awareness of your product. To measure the results, though, you generally have to do market testing, which is an added expense.

If your objective is to motivate customers to buy your product and increase your sales, you can tell how well you're succeeding by your percentage sales increase.

Coming up with a budget

Big corporations spend millions of pounds on advertising and have separate advertising departments. These big businesses spend even more money hiring advertising agencies to convince consumers to purchase their products. If you're a typical inventor, however, you're on a shoestring budget just trying to raise enough money to get a prototype built. Spending money you haven't yet made on an intangible asset like advertising may not seem like a wise use of the little money you have.

We want to assure you that advertising is key to your success. If you don't let people know that you've a great new product for sale, you're never going to sell it. Getting the word out about your invention is essential. In fact, if you map out a good business plan, you take advertising expenditures into account from the very beginning – see Chapter 14 for details on writing your plan.

You can use a couple of different methods to set your ad budget:

- **Percentage of sales:** Even if you don't have any sales yet, you hopefully have projections for the next year, or three, or five years. Allot a percentage of your annual sales, say 10 per cent, to your advertising budget. You can adjust up or down as events warrant, but at least you've a number to start with each year when you work out the overall budget.

- **Cost per customer:** With this method, you use the advertising pounds you spend to determine how much it costs you to sell one product to a customer. For example, if a new product costs £10 and the advertising budget is set at 3 per cent of annual sales, then the cost of advertising for this product is 30 pence.

Spending time upfront crafting your message and choosing your market pays off in the long run. For example, you pay less per ad by signing on to run ads in several issues of a newspaper or magazine rather than paying issue by issue. Likewise, you can save money by having the ad designer prepare a number of ads at once.

If your funds are limited, consider approaching a college or university tutor and offering to let their marketing students design an ad for you as a project. The tutor gets a lesson plan, the students get some real-world experience, and you get an ad for far less than you'd have to pay a professional ad designer. You can even make it a contest and give a small cash prize to the winner.

Supplementing Your Regular Advertising

Advertising isn't always obvious and it isn't always expensive. You have many options when it comes to getting the word out about your invention. You can go the low-key route with promotions or the free route with public relations.

Using alternative methods doesn't mean that you don't have to use regular advertising methods. The methods that we refer to in the following sections are supplements, not replacements.

Pursuing promotions

Promotion is a form of advertising that takes a more subtle approach. Ads shout, 'Come buy me!' Promotions are just a quiet image of your business particulars stamped onto a giveaway or eye-catching item. Some obvious and not-so-obvious places to print your contact information and logo (or other graphic image) include:

- Baseball caps
- Business cards
- Calendars
- Coffee mugs (especially if you're in the food business)
- Delivery vans
- Key rings
- Magnets
- Pens
- Postcards – some to notify customers that a product is available and some to follow-up on a sales call or customer service call
- Sales receipts
- Shopping bags
- T-shirts

Some companies – hundreds of them in fact – specialise in nothing but promotional items. You can find these companies fairly easy as they generally advertise in the back of magazines. You may also try conducting a key word search for 'promotional items' on an Internet search engine.

Your promotional activities aren't limited to just giving things away. Take a look at some other ways to promote your business:

- **Co-sponsor events with non-profit making organisations that advertise your participation.** You can create a tie-in promotion with another business or non-profit making organisation. An example is a local fishing tournament for children in your area, with proceeds going to the Scouts. Your company may co-sponsor the event by donating 25 samples of your new product that the tournament can use as prizes for the kids attending. In return, all the promotional material lists your company as a co-sponsor. Thousands see your company and product names. You're also promoting goodwill in the community, as well as getting publicity.

✔ **Give an educational seminar.** If you need to educate the public about the benefits of your product or show them how to use it, your costs go up. You can sponsor a free educational seminar that's open to the public. The seminar sets you back the cost of the hotel conference room you rent, but if you need to educate your customers, you must bear the costs.

✔ **Sponsor an athletic team.** You can sponsor or co-sponsor a kids' sports team and have your name emblazoned on 11 football or netball shirts.

✔ **Display your product at consumer or business trade shows.** Even if you can't afford a stand, attend the big national shows and possibly international shows if you can afford it, just to keep abreast of what's happening in your industry. And as long as you're there, you may as well pass out your card or some promotional item – everyone else is, so why can't you? (See Chapter 19 for more on trade shows.)

✔ **Develop sales kits with free product samples or application ideas.** Large companies often give away promotional materials – *promos* – or free items at consumer and trade fairs. You can do the same, depending on the nature of your product.

Only your imagination and budget limits the range of promotional tools you can use to deliver your message.

Pushing public relations

Public relations and marketing are generally one and the same department. *Public relations* (PR) is an effort to shape people's opinions about a person, company, or product.

Today, businesses pay attention to public awareness, company name recognition, and product/company loyalty by taking an active role to support consumers' interests and concerns.

Public relations is a visual and verbal image of a company's profile. It creates a sustainable, long-term image, not a short-term glitzy picture. Public relations is management's way of verbalising and visualising a company's mission, beliefs, and values.

Public relations efforts really come into play during a crisis. For example, if a large oil company creates an oil spill, the company is immediately at the forefront and talks about what it does in order to maintain a clean and safe environment. This message can be in the form of paid ads while trying to create goodwill among potential consumers, or it can be free, such as a press conference or time on the evening news. PR means getting your name out to the public in a variety of forms while establishing a positive image for your brand or company.

Table 20-2 compares paid advertising efforts to free public relations. In paid ads, you control what the media says about your product and when, because you provide the advertising material. Unless you pay for the public relations efforts (like the oil company buying television ad space to create a good image after an oil spill), you've little control over what the media says and when. You're at the discretion of the media.

Table 20-2 Comparing Paid Advertising and Free Public Relations

Paid Advertising	Public Relations Promotions
Must be paid for	Free
You determine the message	An editor interprets the message
You control timing	Timing is at the discretion of the media
Intention of most ads is to inform, persuade, or remind about a product – usually with the intention of making a sale	Intention of public relations is often to create goodwill, to keep the company and/or product in front of the public, and to improve the company's reputation
Public reception may be negative – people recognise advertising as an attempt to persuade or manipulate them	Public perceives promotional pieces as neutral or believable
Very powerful at creating image	Can create image, but may not necessarily be what the company wants
Writing style is usually persuasive and effective	Writing style is often creative, and takes a conversational tone, which may inform but not sell

If your company does its best to be positive and create a good image, you can use your creative marketing talents to 'work' with the press and further your company's image for free.

Often, the local media cover success stories on local small businesses. This is free media exposure; you can't control what the media says but you can educate and capture the heart of the journalist to be 'on your side'. The article they present interests readers who are potential purchasers.

Working up some free publicity

One of the easiest ways to get publicity is to write a press or news release. Newspaper and magazine editors are always looking for copy to fill their pages. If you can offer them an interesting story they can write about or a well-written press release that they can turn into an article without making too many changes, you may receive free press coverage for your product. Try writing a one-page story about your new product, invention, or small business success. Don't forget to also send your press release to local radio and TV stations.

Your headline and story must sell themselves to the individual who receives it at the news desk. Think like a general reader and focus on the most interesting or unusual aspects of your invention or accomplishment. The person you send your release to must be able to quickly understand how your new product or business benefits the public. If the benefit isn't immediately obvious, your release goes into the bin.

Get to know the business editor of your local newspaper. Invite them to do a story on you and your product as a special interest article or feature story.

Write your own press releases or articles in the third person (as if you're talking about someone else) and send them to the major newspapers, local publications, and business journals, as well as magazines, trade journals, and other publications in your field. Be sure to answer all the journalistic questions – who, what, when, where, why, and how – and don't forget to include your name and contact information.

A well-written press release can turn your business into an overnight international success story. Take Alex Tew, for example, an inspiring young British entrepreneur who wanted to find a way of paying for socks during his first term at university. His idea was simple: Set up a Web site and sell advertising space in the form of pixels (the small dots that make up your computer screen). He invested in the services of a marketing company to draw up a professional press release; it achieved worldwide exposure for the project and he ended up selling over $1 million of pixels within a few months. Check out his story at: www.milliondollarhomepage.com.

Go to your local library and ask your reference librarian to help you locate books and resource guides that list the names, addresses, phone and fax numbers, e-mail addresses, and individual contacts of the major newspapers around the country. E-mail the news releases out late Sunday night so that your information is on their desks first thing on Monday morning. Follow up the news release with a telephone call on Wednesday.

Some of the most persuasive words used in advertising, whether paid or free, include: Now, free, introducing, starting, revolutionary, bargain, hurry, wanted, quick, easy, magic, miracle, announcing, and improvement.

Creating a run on your product

To spark interest in your product, try the following handy trick. It works like this: The moment your product hits the shelves, you send your family, friends, shareholders, and anyone else you can think of to buy as much of the product as they can. Then, you call the shops and ask if the product is available. Of course it's not available – you just bought it all. If you're really clever, you call the press about the sell-out of your new product.

You create a situation and force the shops to reorder your products, so at least the distribution system gets moving.

Looking at a Few Tricks of the Trade

The following tried and tested advertising hints and tips have worked for others time and time again:

- ✔ Choose a phone number that people can remember. For example: 08000 'INVENT' – which is 08000 468368.

- ✔ Create Web sites for your product and your company. Link them to each other and keep them up to date. Select an easy URL and e-mail address that consumers can remember and list it on all your product's material.

- ✔ Give away freebies, including pens, mouse mats, mugs, and calendars emblazoned with your product or company name, address, phone number, and Web site.

- ✔ Join and attend professional trade association meetings within your own industry – specialise and be the best you can be among your own colleagues.

- ✔ Create shirts with your company's name and logo for your sales team and industry reps.

- ✔ Send follow-up letters to new people you meet at business networking events and conferences.

- ✔ Develop brochures for your new product that use customer testimonials.

✔ Include a postage-paid return envelope or card with your product or company literature in order to get feedback and establish a mailing list.

✔ Publish a newsletter for your customers, buyers, wholesalers, and distributors.

✔ Review non-traditional forms of advertising such as ads on buses park benches, or sponsoring a park bench or local event, with your company or product's name on the event staff's T-shirts.

✔ Code any advertisements and keep track of which ones prove the most effective at bringing in new customers. When you redeem and process the coded coupon, you can tell where the customer got it – a newspaper, a magazine, or postal marketing.

✔ Volunteer in your community at charity events, on planning boards, and at community events – get involved and get your product or company name out there.

Chapter 21

Licensing Your Product

. . . .

. . . .

*A*fter putting in all the hard work of thinking up your invention and then doing the evaluations and testing, handing your baby over to someone else to produce, market, and sell while you rake in the royalties may sound like a good idea. And licensing is certainly a good option for some inventors. However, be warned that only a small number of inventors – about 6 per cent – succeed in licensing their products.

No harm exists in trying, though, and licensing can be the best way to go if you can place your invention with the right licensee. This chapter tells you how to find that ideal licensee and how to license your invention.

Considering Licensing

Licensing is a contractual arrangement in which an individual or company grants specific rights to certain intellectual property. They grant another individual or company these rights in return for royalties and/or another form of payment. Think of licensing as a way of renting your invention to someone else. A *licensee* is a company to which you give permission to manufacture, market, and/or distribute your invention.

This section helps you to determine whether licensing is right for you.

Assessing your chances

If you do your part by investigating the market – choosing potential licensees because they're a good fit for your invention and otherwise doing your homework – you've a much better chance of licensing your invention than someone who's so convinced of their invention's worthiness that they expect the world to beat a path to their door. By putting time, effort, and money into your invention, you better your chances of being part of that 6 per cent licensing success rate.

Licensing companies look for proof (as in market studies) that your product can make money and add to the company's overall product line. Most companies aren't interested in products that

- ✔ are harmful to the environment
- ✔ may cause injury to the user
- ✔ are so totally new to the industry that the customer needs education on how to use them

You may find it more difficult to license into certain industries, including the medical and soft drinks markets, mostly because just a few companies control these industries.

Weighing the pros and cons

Licensing your invention means sharing both the risks and rewards of getting your invention to market. To produce and market your invention on your own you must be able to direct your product from conception to a final marketable package that's ready for the shop shelves, and through all the complicated and costly steps in between. The process costs big bucks, involves lots of risks, and offers no guarantees of success – in fact, you're more likely to go bankrupt than end up flying around in your own private jet! Letting a licensee assume those risks can look mighty appealing.

Of course, the other side of sharing the risks is sharing the rewards, and if you license your invention, you get a smaller proportion of the profits rather than all of them. But sometimes the greater the risk factors, the greater the potential for profit.

Advantages of licensing your invention include the following:

✔ The licensee takes on manufacturing and marketing costs.

✔ You may earn royalties immediately rather than waiting for the break-even point for manufacturing.

✔ The licensee may already sell in foreign markets and can promote your product there.

✔ The licensee's reputation and market position can help sell your invention.

✔ Your invention has the support of the licensee's experienced production, marketing, and sales resources. For example, a large corporation may be able to highlight your invention in a prime-time national television advert – something a small company can only dream of.

✔ Patent infringement is less likely because potential infringers don't want to take on an established company. You may even be able to get the licensee to cover the cost of filing and maintaining the intellectual property.

The disadvantages boil down to having no control over the manufacturing and marketing of your invention and generally earning less money than you would if you reaped all the profits. You also run the risk of having the licensee use knowledge gained from your invention to become a competitor. A risk of this sort is higher with foreign licensees, who may benefit from the different laws and law enforcement in their countries.

Take care to protect your intellectual property by securing the proper patent, design, and trade mark rights. In any interim before you file your applications for a patent or design, ask a potential licensee to sign a confidentiality agreement. Refer to Chapter 2 to find out about confidentiality agreements.

In the end, the quality of your invention, your personal abilities and limitations, and the state of the economy all play a part in determining whether it makes more sense to license your invention or produce it yourself.

Regarding Licensees

You're thinking about licensing your invention – so who may want to license your patent and for what reason?

The *who* may be any number of companies – those that deal with similar or complementary products, those that want to branch out into the niche your invention occupies, or those that produce an item that your invention makes more valuable.

Your invention may be a means of extending the use or life of another patent holder's product, or may threaten an existing product or patent's continued success. Your invention may increase the value of that patent or offer a means to replace the existing product.

The incentive for the licensee is the one that drives the capitalistic system – money. The profit potential depends upon whether the company can develop your invention into something consumers want to buy at a price that makes a profit. In order to make a profit, the cost of your product must be considerably less than the selling price. You have to convince any potential licensee that your product is marketable in sufficient volume and at a low enough cost to make it worthwhile for the licensee to pay you royalties.

Licensees don't license mere ideas. They buy into patented products, or at least those that have a strong chance of getting a patent.

Locating potential licensees

You, better than anyone else, know the intended market for your novel product or unique way of doing things. So you, better than anyone else, know where to start the search for a potential licensee.

- ✔ **The marketplace:** Go where you believe buyers for your invention would look for it. Do your homework and go shopping. Check the packaging of similar items to find out who manufactures and distributes them. Collect catalogues from potential competitors and companies that make similar products. Note the names, addresses, phone numbers, and Web sites of these companies, and then use that information to find out everything you can about what each company does and how well it performs.

- ✔ **The library:** In the reference section of your library, look for business journals that are relevant to your invention and consult the business pages of the national press.

- ✔ **Trade shows:** Chapter 19 has information on finding and making use of trade shows to market your invention. Attending trade shows to meet potential licensees is an absolute must.

- ✔ **The Internet:** Conduct a search by type of product on the Internet to find many more companies. You can also type in the Web addresses of companies you find as you investigate the marketplace and attend trade shows. And while you're online, check out the UK Trade Association Forum Web site (www.taforum.org), which holds a national directory of trade associations and business sectors.

- ✔ **UK Intellectual Property Office Web site (www.ipo.gov.uk):** Running patent searches on products similar to yours can tell you who the licensees are for patents.

One of the least productive means of locating potential licensees is approaching the research and development departments of multinational companies. Their job is to develop new products for their company, and they tend to think that if they didn't invent it, their company doesn't need it. Or they don't like to give their managers the opportunity to say, 'This is a good idea. How come you lot didn't think of it?' We refer to this outlook as the 'not invented here' syndrome.

Considering potential licensees

Your first consideration regarding a potential licensee is the company's financial stability. You want to get your invention in with a company that has a proven track record and a foreseeable future. Also make sure that your potential licensee is good at marketing the products or methods it currently holds the rights to. And although many companies can manufacture your invention, a good licensee also has a distribution system in place.

Put some time into finding out about the companies you consider. Look at the *operational profiles* of the industry they're in; that is, find out who the major retailers, wholesalers, suppliers, and sales reps involved in that industry; the average costs of bringing a product to market in the industry; the general market climate; and recent history and projected trends for the industry. Find out about each company's sales numbers, product lines, and how it sells products. Research any lawsuits the company faces.

Along with the sources you use to find potential licensees (see the 'Locating potential licensees' section, earlier in this chapter), you can use Companies House to access general company information, including filed accounts and annual returns. Visit the Companies House Web site for more information (www.companieshouse.gov.uk).

If you research a public limited company (plc) you may be able to download recent annual reports from the company's own Web site. Annual reports contain such valuable information as:

- ✔ When the company started.
- ✔ How much money the company makes and percentages of sales by product sector.
- ✔ The company's marketing techniques – whether it sells directly, uses distributors or independent sales reps, or some combination.
- ✔ How much the company pays in royalties and when.
- ✔ How much the company spends on advertising.

- ✔ The company's production capabilities.

- ✔ Any legal action the company may be involved in.

- ✔ Projections of future revenues.

- ✔ Product development information – how much the company spends on prototypes, moulds, packaging, intellectual property, and so on.

- ✔ Information on competition, market share, and sales – often presented in a pie chart that shows the company's and competitors' slices of the market.

With this basic information, you can start to approach both manufacturers and distributors to gauge their interest in licensing your invention.

If you license your invention to the biggest retailer on the planet, be aware that the company, not you, sets all the terms. The giant tells you the price it's prepared to pay and that price probably won't be up for negotiation.

Researching companies before you approach them saves you time, aggravation, and the pain of rejection. You waste time and look unprofessional if you approach companies without knowing whether your invention is compatible with the company's outlook.

Assume that you find 15 companies that you think would be a good fit for your product. After doing some research, you realise that four of those companies control about 80 per cent of the overall market. Your best bet is to approach each of those four companies.

Contacting Potential Licensees

You're ready to start contacting the companies that you determine seem best suited to making the most of your invention. Your basic format must create immediate interest for the potential licensee, in terms of what they need to do before they can package the product and get it on the shop shelves or in the catalogue.

Gathering the materials you need

One of the reasons you want to license your invention is because you'd rather leave all the market research and selling to someone else. The bad news is that you have to do both those jobs to sign with a licensee. You must put together a convincing package to market yourself and your product. You need to provide reliable information that includes:

- ✓ **A description:** List what your invention is, what it does, and how it does it. Point out your invention's unique characteristics and describe why and how it's better than anything similar that's currently on the market. Tout all its various applications.

- ✓ **Proof that your invention works:** Just because you've a patent doesn't necessarily mean that your invention works. Include sketches, samples, and a prototype if you have it, as well as engineering specifications and tooling and manufacturing requirements.

- ✓ **Information on intellectual property:** Let people know where you've got to in the process of acquiring patents, registered designs, trade marks, and so on.

- ✓ **Market analysis:** Provide documented material and research that shows that a market exists for your product. Chapter 18 has the low-down on conducting market research.

You also need to be prepared to state what you expect from the company in terms of royalty arrangements and what type of licensing agreement you want. You don't get into the details of the agreement at this point, but you can sketch the broad outlines of what you seek. Chapter 22 talks about negotiating an agreement and the 'Examining Types of Licensing Agreements' section, later in this chapter, provides lots of useful information.

Make sure that all the information you provide to potential licensees is accurate and as brief as possible. If you're just setting up a one-to-one appointment to demonstrate your invention in person, you don't need to go into detail about how your invention works; you focus instead on how it fits into the company's product line.

Making your way to making your pitch

The traditional way of contacting potential licensees is to send a letter to the director of new product development. The letter introduces your product and explains how it can fit into the company's product line and increase market presence and profits. Typically, an approach on these lines gets you a rejection letter, if you get any response at all.

If you get an interested response, your next contact is likely to be with someone in the product development department. You're by no means on the home run. In fact, a high probability exists that the company meets your idea with scorn and disdain, because the people in the research and development area didn't think of it themselves.

The best way to approach potential licensees is to communicate directly with the key decision makers. Even better is enlisting a key decision maker as a champion for your invention. A *champion* is a person who likes your product, understands its value, and is willing to work with you in a mutually beneficial relationship to bring your product to market. A great licensing agent who knows every contact in the industry can be a champion, as can the boss of a manufacturing company who can see the money-making potential of your invention for his company.

Joining related trade organisations is a great way to gain an 'in' to the industry and is an invaluable resource for networking and gaining inside information on the industry you want to break into.

Considering the company's side

Large companies come under constant attack in our increasingly litigious society. Not only do consumers sue them, but so do their employees, other companies, and – most pertinent to this discussion – disgruntled inventors.

Consider the following cautionary tale: An inventor approaches the fictional Grill Accessory Company about licensing his new grill gadget and gets turned down. A couple of months later, the inventor sees that the Grill Accessory Company is selling a product very similar to his. The inventor hires a lawyer, who slaps a lawsuit on the company. Now, a little common sense would tell the inventor and his lawyer that the similar gadget was in the pipeline long before the inventor approached the company. It takes months and months of preparation to get a new product to market – nothing goes from concept to market within a couple of months. But still, the Grill Accessory Company's lawyers have to defend a lawsuit they know that they'll win.

Most large corporations want you to sign their confidentiality agreement before they review your product, and they often resist signing any confidentiality agreement that you provide. Some companies won't even talk to you until you file a patent application that clearly defines your invention.

Get sound legal advice to help you deal with such a situation. Signing another company's agreement can give that company free rein to blab your product idea or to keep it secret for just six months or a year. An agreement may prevent you from showing your product to other companies until the first company is ready to turn you down. If you're not careful, you can give your rights to another company to further develop a product based on your idea and go to market with it. Some agreements may not only take an inventor's current idea in a licensing agreement but also ownership of the rights to all future ideas!

Many paranoid inventors think that all large companies are in the business of ripping them off. And, though some companies are certainly not on the up and up, more often the lone inventor tries to take advantage of a company than the other way round.

Facing Rejection

Companies reject invention submissions for new products for several reasons. One of the primary reasons is the 'not invented here' syndrome. Large companies have in-house research and development personnel that include engineers, scientists, and product designers, whose job is to develop new ideas into products. Often, it makes more sense for the company to depend on its employees to develop appropriate products than to risk placing the company in a liability position. A company that relies solely on its own inventors also saves the time and money it takes to review and evaluate new products. To find out information regarding a company's standard policy in working with inventors, contact the company directly and ask for information.

If you get turned down, make sure it's by the right person in the company (a key decision maker) and not by a receptionist or warehouse assistant.

Licensing to your own company

You've heard of having your cake and eating it too. Well, here's the happy situation that can be yours if you can license your invention to a company that you control.

If you're a serial inventor (not to be confused with a serial killer) who hit it big with an earlier invention, you may have formed your own company and retained a controlling interest. In this case, you've a ready-made platform for launching subsequent inventions.

Financially, your picture couldn't be rosier. You now have several income sources – a salary as

managing director with promotional and travel expense accounts, royalty payments, and potential share dividends and increasing share values.

The financial arrangements of licensing to your own company can get a little tricky because you have to keep the company's books and interests balanced. You may need to adjust share holdings or move to a different corporate and legal structure. Seek sound financial advice from your accountant or financial adviser before striking a licensing agreement with yourself.

Say you got to make your pitch to the appropriate people in the company (and you really thought they'd go for it), but the company turned you down anyway. The company may have rejected your invention for any number of reasons, and very few of them relate to the merits of your invention. Some of the reasons that companies don't license inventors' products are:

- **Product related:**
  - The invention doesn't fit into the company's overall market. (You can prevent rejection on these grounds by doing your homework on the company's product line.)
  - The product doesn't meet government standards or is environmentally unfriendly.
  - The product life cycle is too short.

- **Financial:**
  - The company isn't in a position to pursue outside inventions at this time.
  - The invention doesn't have a large enough profit margin.
  - The invention requires too much long-term investment – in infrastructure or research and development, for example.
  - The cost of maintaining *ISO standards* (international standards designed to level the playing the field in different countries with regard to quality systems) and other regulations is prohibitive.

- **Market related:**
  - The company tried to market similar products that didn't sell.
  - The market for your invention isn't big enough to make producing it worthwhile because the sales won't be enough in the long run.

- **Legal:**
  - Your product has limited patent protection.
  - The company is concerned about liability issues relating to your product, including safety, and other concerns.

A company may reject your invention for a reason you may never hear about directly – the inventor is too difficult to work with. If the company thinks dealing with you may cause more headaches than profit, it can, and will, turn you down.

Examining Types of Licensing Agreements

Licensing agreements come in three basic types: *exclusive*, which gives just one company those rights, even to the exclusion of you yourself; *sole*, which gives just one company the rights but lets you retain the option of also exploiting the rights; and *non-exclusive*, which grants more than one company rights to one or more aspects of making and marketing your invention. The following sections talk about each type of agreement in turn.

Being exclusive

An agreement to license your invention can grant the licensee exclusive rights to your product and/or method, which means that one company has total ownership. Technically, this one company has permission to undertake or control the entire manufacturing, sales, and distribution process from Glasgow to Moscow.

An exclusive licensing agreement has the advantage of simplifying things – you have one relationship with one business. One company undertakes or controls the manufacture, marketing, and distribution of your product.

Licensing agreements are complicated legal documents and, as with any legal document, seek the advice of a knowledgeable lawyer before signing. You can check out a sample exclusive licensing agreement in Appendix A.

Being not quite so exclusive

Although you may want to license to just one big established manufacturer, Company X, the task of exploiting the major worldwide market for your product, you may like the idea of continuing to make and sell the invention in your own chosen area. You may be an established small manufacturer supplying a local area, and you want to continue doing so while the big companies in Company X hit the overseas markets. A solution here can be a sole licence. Company X gets the big market, defined geographically or by numbers of units produced, while you continue making and selling the improved and patented version of your product.

One of the appeals of the sole licence arrangement, both for you and Company X, is that you can maintain the good relations you've built up with your local customers and continue to develop further improvements in the product. The agreement with Company X needs to address the question of such product improvements: whether you included them within the original agreement and remuneration, or whether you negotiate further terms that make them available to Company X, exclusively or otherwise.

Being non-exclusive

A non-exclusive agreement means that you can sell your product to more than one company. You can make a fortune by allowing several different companies the rights to use your new process or product. And you can also keep making and selling the product yourself if you so choose.

If you plan to manufacture your own product, you may want to give the distribution rights to several companies. You can give a major distributor the British rights and split up foreign distribution rights to companies in those countries.

Understanding how these agreements work for you and the licensee

You can sign non-exclusive agreements with several distributors throughout the country or world or sign agreements with companies in different markets or aspects of the business. For example, you can license one company the rights for the mail-order industry, another company the rights for television, and another company the rights to sell in retail outlets. By doing so, you can make money from each specific market instead of concentrating on just one.

If you decide to grant a non-exclusive licence for your invention you must consider a number of factors. If your invention is a commodity with extensive manufacturing costs, potential licensees want to know that if they spend the money on the start-up manufacturing costs, they won't have competition in the same market to sell and distribute the invention. The small, non-exclusive licensee worries that a larger company with more financial resources can develop better tooling to lower manufacturing costs, thus lowering the sale price of the product and/or increasing the large company's profit margins. To deal with these fears, non-exclusive licensing agreements often assign trading areas that give a company exclusive rights within certain geographical areas.

If your invention consists of an add-on to another invention, or a widely used product with no patent that a number of companies manufacture, a non-exclusive licensing agreement may be the best deal for all concerned. So say you invent a device that increases the efficiency of breathing machines used in hospitals around the world. Nine manufacturers produce the breathing machines, and each one is interested in your add-on invention, because it improves each machine. You can license your invention to each company that makes the breathing machines. Each brand remains competitive and holds onto their existing market, and you sign nine licensing agreements and collect nine royalty payments.

Looking at ways to use non-exclusive agreements to your advantage

You can license your product for sale in different media. For example, you may have one agreement with a television home-shopping channel, another agreement with a distributor that supplies retailers, and a third with a catalogue company that sells through mail order.

You can also license by geographic location and define the geographical areas however you want. Setting out precise areas can remove doubts and settle fears the licensee may have over paying you a licensing fee and then realising no profit because many others in the area also sell your product.

You can license by services rendered. For example, one company has a great distribution system and sells products at a national, as well as international, level, but doesn't have a manufacturing plant. You can give marketing rights to that company and manufacturing rights to another company.

Taking Care of Foreign Licensing

Licensing your product to a foreign company requires a carefully crafted licensing agreement. Consulting a lawyer is important because rules differ from country to country – and your UK patent is good only within the UK. Be careful that the agreement doesn't violate host country anti-trust laws. Under the anti-trust laws of many countries, the licensor can't set the price at which the licensee resells a product.

Bear in mind that if you negotiate a deal with a manufacturing company in China, for example, you may just be sent contracts written in Chinese! Seek the services of a professional translator before signing on the dotted line, to make sure that the company didn't amend the agreed conditions.

At times, because of import and export restrictions between trading countries, a wise move is to select reliable manufacturing marketers to make and distribute your invention in their trading areas. Coping with the differences in patent law in each country (as well as local customs and regulations) is a challenge even for experienced multinational organisations.

Chapter 22

Negotiating a Licence

. . . .

. . . .

*I*f you decide that licensing is the best route-to-market for your invention (see Chapter 21), the construction of the licensing agreement is the key to a profitable mutual relationship.

A licensing arrangement can be a long-term commitment, so you want to take time before you agree the deal to be sure that you want to associate with this company and these people for the long haul. And remember that any company worth having such a relationship with has similar questions about you.

This chapter goes through the various stages of the negotiations and offers tips for achieving a successful and equitable licensing contract.

Employing the Art of Negotiation

The fine art of negotiation relates closely to the fine art of compromise. A negotiation starts when one party makes a bid to obtain something from another party. In the back and forth of negotiation, each side makes concessions in an effort to seal the deal. Almost all agreements end up as a compromise between the negotiators.

In all likelihood, your negotiation session with a potential licensee feels as though you're David taking on Goliath. Try not to feel intimidated by the prospective licensee's team of lawyers, engineers, scientists, and marketing analysts who confront you – all focused on obtaining the best bargain for their employer. Meanwhile, your focal point is bartering a deal that supports your interests. Somehow, together, you must forge an agreement that serves both parties' goals.

A successful negotiation ends with both you and the licensee feeling that you achieved a fair deal, which creates the foundations on which you build an amicable long-term working relationship. Getting along after the deal is done serves everyone's best interests. You can be a great asset to the company from a consulting standpoint, and the company may want to have first right of refusal on additional products that you invent.

The following sections discuss the key points in the negotiation process and show you how to make them work for you.

Submitting to an evaluation

If you don't get a rejection letter straight away, consider yourself lucky. Most potential licensees reject inventions within a short period. But even if you're fortunate enough to be one of the lucky few picked for further review, don't start counting your remuneration yet. Any company considering licensing your invention wants an opportunity to look at your product from every angle in order to determine whether it's interested in pursuing talks with you.

The company investigates your product, examines its own resources, and looks for answers to the following questions:

- ✔ Does your invention truly have a market and how big is that market?
- ✔ Do we have the equipment and staff to manufacture and market this product effectively?
- ✔ What are the estimated costs for manufacturing, marketing, and distribution?
- ✔ How is the competition going to react to this invention?
- ✔ How much time and effort do we need to invest to introduce the product to market?
- ✔ Is the risk is worth the reward? Does this invention have the potential to make enough money?

During the examination, the company may also explore potential marketing strategies and determine whether it has the facilities to service your invention.

Before you turn over your invention to the potential licensee for evaluation, make sure that you have your patent application(s) in place or, if that's not possible, get a signed confidentiality agreement. See Chapter 2 for more on confidentiality agreements.

While all this scrutinising is going on, the potential licensee may not want you to show your invention to anyone else. That can mean weeks of losing opportunities to present the invention to other prospective licensees. For this you deserve compensation. Before you turn over your invention for testing, ask the company for a *letter of intent* that details the specifics about the evaluation period – how long it is, what tests they may and may not conduct during the evaluation, and how much your compensation is. Figuring out what to ask for in compensation is tricky. Basically, you try to determine what the period in which you're not showing your invention to anyone else is worth to the evaluating company. If the level of interest is high, you can ask for more than you can if your invention makes a nice addition to the company's product line but isn't a big money-maker.

Allowing this evaluation is generally in your best interests. After all, do you really want a licensing agreement with a company that doesn't do its homework? Worse yet, do you want to find out, after you sign an agreement, that the company can't get the product to market? The latter is particularly bad if the agreement bases your royalties on a percentage of profits generated by sales.

Nonetheless, permitting the evaluation involves acceptance of certain risks. For example, by scrutinising your invention, the examiners may discover how to make a better product. They can then choose not to license your invention, go to market with their own product, and never pay you a penny beyond the evaluation compensation. Horror stories about such happenings circulate throughout the invention community. Some inventors become so distrustful of companies – so fearful that someone may steal their inventions – that they never show their inventions to anyone. Obviously, such overprotection doesn't benefit the inventor. Instead, let common sense prevail in determining who sees your invention and when. Until someone who can produce the device sees it, you've no chance of realising the potential of your brainchild.

Whether or not a company decides to license your product after evaluating it, seize the opportunity to find out what the company likes about your invention and what it doesn't like. Especially if the company ends up not licensing your product or process, knowing what put them off can help you fine-tune your invention for the next suitor.

Preparing for the meeting

Having your paperwork in order and your presentation polished is key to getting what you want at the negotiating table. Remember, this is potentially a big money deal. It makes sense to show that you consider your invention worthy of significant consideration by being prepared to negotiate. This is crucial.

Gather up-to-date prototypes, drawings, and engineering notes. Likewise, be able to back up your sales projections with numbers from your market research (see Chapter 18), and be ready to produce credible production-cost estimates.

Be ready to answer the following questions:

- ✔ What type of consumer does your invention appeal to?
- ✔ How many of these buyers are there?
- ✔ How much are consumers willing to pay for the product?
- ✔ How much is your invention going to cost to produce?
- ✔ What are the projected sales figures for the first year?
- ✔ What does the competition offer and how does your invention beat their products?
- ✔ What are at least three significant differences between your product and that of your competition?

Be prepared to back up your answers with credible data. If you can get statements of intent to buy from potential customers to back up your sales projections, do so. Documentation that supports your projections of your invention's market value can be very helpful to your cause – particularly if the author is a professional marketing firm or a recognised authority in the field.

Having such factual information shows the licensee that you aren't in the dark about the potential worth of your product and helps establish how reasonable your payment demands are.

Do your best to anticipate the questions or issues that the company may bring up in the meeting, and devise a strategy for dealing with them. Walk through your worst-case scenario and come up with ways to diffuse a negative situation and keep the discussion on a positive path.

In the same vein, try to anticipate concessions for both sides. Run a reality check on the feasibility of the requests you plan to make of the company, and try to figure out whether you want something that would cost the potential licensee little to give up. Recognise that the company you negotiate with spends time contemplating what your requirements may be and what trade-offs they may make to conclude the negotiation in their favour.

Thinking ahead pays off during the actual negotiating situations. Such in-depth planning for the meeting has several benefits:

✔ Your confidence in your ability to clinch the deal increases.

✔ You may avoid making hasty decisions that you regret later.

✔ Well-thought-out offers and counter-offers enhance your image and credibility at the negotiating table.

In addition, talking through what's important to you with your accountant, business partner, or advisers ahead of time clues you in to what considerations are important to you.

And you need to be emotionally prepared. Before going into the meeting, get psyched up. Mentally review all the reasons the company should license your product – and why you're the person to convince this company to give you a contract.

Laying out the agenda

Reaching the negotiating table means that you've got a lot right, but things definitely become more challenging now. You must effectively handle the rigours and pressures of bargaining in order to walk away from the negotiations as a winner.

If you don't have a patent yet, you and the prospective licensee should sign a formal document that specifically identifies the intellectual property you intend to discuss – before the real negotiations start. This agreement provides some protection for your invention for a stated period of time.

You may be tempted to require the licensee not to challenge your intellectual property rights, but check with your legal advisers whether a no-challenge clause would breach competition law in the licensed area.

Sorting out stress

Some negotiators thrive on stress at the bargaining table, finding the adrenaline-charged atmosphere keeps their minds alert and able to react quickly to new proposals. For those not overly familiar with negotiations, the reverse is often the case. The stress messes up their mental processes and hinders clear thought.

If the very idea of negotiating brings you out in a cold sweat, you can relieve a lot of stress by preparing well and anticipating issues and answers. You can also help reduce stress in the moment by exploiting natural breaks in the meeting to relieve the tension and reflect on the next points to cover. Some negotiators even resort to relaxation techniques before a meeting, during a break, or even during a negotiating

session. If you're one of these people, try the following:

✔ Take three or four slow, deep breaths. Hold each breath in at the top for a few seconds before slowly exhaling all the way.

✔ Mentally and physically relax your muscles. It may help to clench everything for a few seconds, and then let go all at once. Or, you can work your way up your body, mentally telling each part of your body to relax.

Periodically tell yourself that you're doing fine, that everything is okay, and that the process is going forward just as it should. And remember that the other side may be just as stressed as you are!

Every deal is different but the overall goal is the same: You want to obtain a fair price from the right company to produce your product. Your remuneration typically includes an initial lump sum payment, followed by ongoing royalties. The important issues you need to address to get to that goal are:

✔ **Negotiating how much money you get upfront:** See the 'Getting reimbursed upfront' section, later in this chapter.

✔ **Determining the royalty rate and how the company computes it:** The 'Researching royalty rates', later in this chapter, has more on this.

✔ **Setting a minimum annual royalty guarantee:** The 'Receiving a guaranteed income' section, later in this chapter, covers this.

✔ **Arranging the commencement date and dates of royalty payments:** Typically, the company pays royalties quarterly.

✔ **Limiting the length of your licensing agreement:** State how long the agreement is good for.

✔ **Defining the conditions under which either party can terminate the agreement:** Even if you originally define a length of time for your agreement, you may want to include an option for getting out early – and any penalties for doing so. You may want the conditions to cover what happens if the licensee company goes bust or undergoes a take over.

✔ **Laying out the terms of the default clause:** See the 'Receiving a guaranteed income' section, later in this chapter.

✔ **Examining the terms and obligations the agreement binds you to:** Chapter 21 goes through the different types of licensing agreements.

Make sure that you and/or your accountant have the right to audit the accounts of the company you license to.

The potential licensee wants to know what you think in relation to royalties and a licensing agreement, to make sure that you're both in the same ballpark and negotiating is worthwhile. From a negotiation standpoint, it's better to have the licensee make an offer to you after you sell them on your product. That way, you can see what they're thinking and counter-negotiate. When a potential licensee asks what terms you're looking for, a good response is, 'My terms are negotiable. What I want to do is to create a win–win situation for both of us. What do you think can work best for you, given your past record?'

Be very careful if the licensee wants to re-negotiate your agreement. The company is probably trying to get more favourable terms for itself, and any modification of the existing agreement may automatically and legally cancel all or part of the original agreement.

Regarding your bargaining position

Your bartering power with a potential licensee depends upon how badly they want access to your invention. Your greatest strength comes if you have a patent for an invention that adds value to – or threatens – an existing product. Your next-best position is having an invention that complements the primary product line of a manufacturer or marketer.

Aside from these two situations, how well you present your invention to a manufacturer or marketer in terms of earning power dictates your negotiating position. In order to make a convincing case, you have to provide the market data to support your claims.

Negotiating like a pro

The art of professional, hard-nosed negotiation is one of careful determination. Each side wants to see how much the other side concedes without terminating discussions. Each side must be aware of the other's position and attempt to bring the sides together with acceptable compromises.

Keep the following in mind as you negotiate:

✔ **Be ready to walk out if you must.** Set your realistic minimum requirements ahead of time and be prepared to end the talks if it becomes clear that the company isn't willing to meet those minimums.

- ✔ **Don't reveal your bottom line too early in the negotiations.** The company may offer more than you anticipate.

- ✔ **Don't walk out too quickly.** Don't use any little excuse to stop the negotiations. After all, you want to license your invention, and you've got to this point with this company, so give the process a chance.

- ✔ **Don't enter the strategic manoeuvring arena alone.** Unless you're particularly skilled and experienced in negotiations, take along someone who can help level the playing field, such as a business partner or adviser with experience in negotiating licenses or commercial agreements. Legal issues abound in any licensing agreement and unless you've done this before, you're a fish out of water and may be finished in no time.

- ✔ **Don't take the negotiations personally.** Don't interpret questions about your invention as criticisms, and don't take criticism of your invention as a slur upon your character. Remember that this is business.

- ✔ **Keep your emotions in check.** Don't get cross and don't be defensive. Negotiating calls for calm, cool, and collected behaviour. If this isn't you, get someone else to represent your interests at the negotiations. Likewise, rein in your glee. You may have to control that high until . . . well, until you receive your first remuneration.

- ✔ **Give reasons for your responses.** If you say no to a proposal, explain why. If you disagree strongly, say so – nicely – and propose a compromise if you can.

A serious and realistic inventor accepts failure as part of the inventing process. Treat rejection during the licensing process as just another problem to solve. Analyse why you can't get together with your target licensee; perhaps you two were a mismatch from the beginning. Often what hurts the most is reaching a point in the transaction where all you need to do is finish up a few minor details, when suddenly a major obstacle springs up and kills the deal. Even then, make your motto 'Try, try, and try again'. Tomorrow is another day.

If the deal seems too good to be true, it probably is. Get good sound legal advice before signing any licensing agreement.

Estimating Your Invention's Value

In order to negotiate a licensing agreement, you and the licensee have to establish a mutual value for your invention. This is the first issue in a series of negotiations.

All inventions have value – at least to the inventor. However, the question isn't how much value you place on your invention, but how much value other people perceive your invention to have. Value, like beauty, is in the eye of the beholder.

As the inventor, you're naturally inclined to assign a high worth to the product of your inspiration and perspiration. However, if you want to license your product, you need to live in the real world, where your invention's worth only as much as the profit it can make. So, keep your ego and your wildest dreams in check while you're at the bargaining table.

Back in the real world again, perceived value is a little hard to measure, which is why you need to go to the negotiating table with some real facts and figures at your fingertips. You want to show potential licensees estimated production costs and the results of studies that indicate how much customers are willing to pay for your invention. For more on figuring out how much customers are prepared to pay, take a look at Chapters 18 and 19.

At this point, all anyone can do is *guesstimate* (a not-too-technical combination of *guess* and *estimate* that sounds more reliable than plain old *guess*) costs and sales. But the better the numbers look to begin with, the better the chances that both sides are happy down the road when guesstimated numbers come close to actual figures.

Be aware that guesstimating is an imprecise science. The only thing you can do is analyse the industry, look at the sales and market share of a similar product selling within the industry, and project growth rates based on industry averages. You use existing information as a base for your guesses. Certainly, an experienced manufacturer can make educated guesses as to materials and production costs, but sales volume affects these and the guesses may turn out to be off the mark.

Many agreements use guesstimated numbers with provisions for altering the terms in line with actual figures. Keep in mind that getting people to buy a new product necessitates change in their behaviour. As a species, humans resist change, and overcoming this resistance to get people to buy your invention costs time and money. The licensee has to bear the expense of educating consumers about your invention and how to use it.

Contemplating Compensation

Getting paid for your work is what licensing is all about. From your perspective, you've already done the hard part; now it's time to sit back, make yourself comfortable, and wait for the royalties to come flooding in. Of course, few of us inventor-types can name a price and have companies begging to meet it, so you have to get your head out of the clouds long enough to figure out how to negotiate realistic and fair compensation.

You can value inventions in a variety of ways. In most situations, the major items that you review to come up with the estimated value of your invention include sales of similar products, the profit margins, estimated market share your invention should maintain in a competitive environment, who the competitors are, how far the product merits a patent, how strong the claims on your patent are, and the long-term projected growth rates. Again, an important factor here is how well you or your designated negotiator can negotiate.

Be creative about how the company compensates you. Consider including share options, or receiving a certain amount of money per item sold or per use, or whatever other creative scheme you and the licensee can come up with. Most important, remember that everyone involved needs to profit, or no one wants to sign an agreement.

Don't get greedy! Many deals have been ruined by greedy inventors.

Getting reimbursed upfront

In addition to the actual royalties, many licensing agreements include the licensee (the company) reimbursing the licensor (you) for money already spent or in compensation for time. For example, if your invention requires a long development and manufacturing period, you can include a provision that provides you with some income during the introduction period.

So a typical licence agreement includes an *upfront fee*, which is the payment the licensee makes, possibly in stages, to help you recoup some of the costs you put forth in doing research and development, conducting market studies, and obtaining intellectual property protection. And the agreement should also reflect the often considerable know-how and perhaps trade secrets that you must hand over to get the licensee started.

Many inventors assume that this fee is comparable to winning the National Lottery. Sorry to burst your bubble, but upfront fee amounts vary by product and industry. For example, though it can take millions of pounds in research and development costs for a new type of medical device or drug, the same costs for a new screwdriver may be minimal.

For a consumer-orientated, mass-merchandise product, the licensee typically pays between £5,000 and £50,000 in upfront fees. The fee depends on the product, the industry, how far advanced you are in your product's life cycle, how much you had to spend to get your product to that point, and how well you can negotiate.

Receiving a guaranteed income

No matter what the rest of the agreement ends up looking like, try to negotiate getting a guaranteed minimum payment that isn't dependent on sales or production numbers. Also include a non-payment clause in the agreement that specifies what happens if the licensee doesn't pay up.

Especially if your invention threatens your licensee's product, you run the risk of licensing your product or process to a company that has no intention of ever marketing it. For example, say you invent puncture-proof vehicle tyres. You know that you have a sure-fire winner, because drivers are happily going to pay a premium for a product that ensures that they never suffer a puncture or have to pay for replacement tyres. You negotiate with a company that currently has the largest portion of the tyre market. The exclusive, life-time agreement offers you a 10 per cent royalty on every puncture-proof tyre that the company sells. You jump at the offer because you know that the industry standard is closer to 5 per cent, and your tyres can sell for a bit more because they last longer than regular tyres. You go home and wait for the grand announcement of the revolutionary new system. And you wait . . . and wait . . . until finally it dawns on you that the announcement is never coming.

See, the tyre company may have no interest in (and no intention of) replacing their product that drivers buy frequently with your product that results in drivers buying replacements less frequently. And not only is all the time and effort you put into inventing and testing the system down the drain, the exclusive, lifetime agreement you signed means that your invention is, to all intents and purposes, dead. If the company never offers the tyres for sale, no one ever buys them, and 10 per cent of nothing is nothing. Without a mini-mum performance guarantee clause, the licensee can sit on your invention while you starve to death. Don't make this mistake!

Insisting on a default clause

The way to avoid this pitfall is to make sure that you get a payment upfront and that you include a *default clause* in the agreement that forces the company to actually produce your invention. Essentially, the default clause says, 'Look, manufacturing company, you say you can manufacture, market, and distribute my product for me and I trust you to do so; now, if you go back on your word, it's not my problem – it's yours. After all, I gave you – and not anyone else – the rights to make and sell it and that costs me money; there-fore, you guarantee that you'll sell x number of units per year or pay me x amount of money regardless. It's not my fault if you can't sell.' Including a default clause is a must.

Also, think twice or even three times about granting a company a lifetime, exclusive licensing right to your invention. Your goal as an inventor is not to have the licensee pay you a minimum guarantee to keep your product or methodology out of the marketplace in order to prevent competition with another of its products. Nor do you want the licensee to be satisfied with a minimal licensing return because of a limited distribution system.

Steering clear of balloon payments

Don't enter into agreements with *balloon payments*, which means getting all your money at the end of the agreement term. The company may not be able to pay you by the time your money is due. The licensee may make money selling your invention, but lose money through bad management or any other reason, and you never get a penny. Ask for quarterly payments, but you can also negotiate for shorter payment intervals.

Including a non-payment clause

A *non-payment clause* states that if the licensee doesn't pay royalties to the inventor for a specified length of time (usually a quarter), you can cancel the agreement. A non-payment clause should stipulate that the licensee pays a cancellation fee equal to three to five times the annual minimal payment, stops marketing the invention, and gives any unsold products to you. Insist on a non-payment clause as well.

Researching royalty rates

A *royalty* is a share of the proceeds paid to the owner of intellectual property in compensation for using the property. The royalty can be a lump sum payment, a set amount per year, a percentage of sales, or a fixed amount per item sold.

Watch the wording of your agreement for everything, but especially when it comes to the royalty schedule. You may be hoodwinked by terms such as:

- **After discount:** Suggested selling price less the percentage discount to the wholesaler or distributor.

- **Wholesale:** The price at which the manufacturing company sells a product to a wholesaler or distributor.

- **Recommended retail price:** What the manufacturer considers to be a fair market price for the consumer to pay.

- **Retail:** The price the product actually sells for.

> ✔ **Sales promotion discount exceptions:** The additional discounting of a product to wholesalers and distributors. Manufacturers use additional discounting exclusively for sales promotions.

> ✔ **Lot quantity:** A discount on a volume basis. The higher the volume, the bigger the discount.

If you license your invention to a manufacturer outside the UK, the company may want to pay you royalties in a currency other than pounds sterling. If this is the case, check out the exchange rate between the pound and the other currency and the all-too difficult predictions on how rates may change. Fluctuations in the exchange rate can have a dramatic effect on how much you think that you'll receive and how much you actually receive.

The typical royalty rate is generally between 2 and 10 per cent. Inventors commonly cite 5 per cent, which means that you get 5 per cent of the money received by the manufacturing company when it sells your invention. But royalty rates can vary from 0.01 per cent to 20 per cent, depending on the product, the industry, the intellectual property rights that you hold, and how well you can negotiate. Table 22-1 lists typical royalty percentage ranges that inventors have been able to achieve for various industries that inventors commonly accept.

Table 22-1	Typical Royalty Percentages
Industry	*Royalty Percentage*
Automotive	5–9
Electric/consumer	6–9
Hardware/housewares	5–9
Machining	5–8
Packaging and plastics	5–9
Plumbing/heating	6–14
Retail–consumer	6–12
Sporting goods	7–13

Source: Based on data from Inventor's Handbook

The more developed the technology, the higher the royalty rate the inventor receives. The average royalty may be around 5 to 7 per cent, but it varies greatly from field to field. Technologies requiring lengthy and expensive regulatory approval may reap lower royalties than less complicated inventions. Yet royalties on high margin products, such as farm machinery, can range up to 20 per cent and royalties on software can range up to 35 per cent.

Yet royalties on high margin products, such as farm machinery, can range up to 20 per cent and royalties on software can range up to 35 per cent.

Some royalty schedules include a flat fee per unit sold. Others call for sliding fees, which we demonstrate in Table 22-2.

Table 22-2	Sliding Royalty Schedule
Number of Units Sold	*Royalty Percentage*
1,000	10
1,001–10,000	8
10,001–100,000	5
100,001 and over	2.5

Don't allow the licensee to insert in the agreement that the licensee does not have to pay royalties until it receives payment from customers. The licensee's credit arrangements are its business, not yours.

Don't expect the manufacturer to pay royalties on items that customers have returned. The manufacturer bases your royalty payments on items sold, not items shipped. Customers may return a product to the shop, or the manufacturer may recall the product for some reason, and your royalty payment reduces because of these losses.

If your invention is part of a sales promotion and the manufacturer sells it at a discounted price to wholesalers and distributors, your price-based royalty payment reduces to reflect the discounted price.

Part VI
The Part of Tens

'You've no idea what it is but you'd like it patented anyway?'

In this part . . .

It could hardly be a *For Dummies* book without a Part of Tens, now could it?

These short and sweet chapters list helpful organisations for inventors, and include brief but inspiring sketches of successful inventors and inventions that had an impact in the past – and continue to do so.

Chapter 23

Ten Key Contacts

. . . .

. . . .

*T*his chapter lists a range of organisations that can help you get your invention off the ground.

Anti Copying in Design (ACID)

After other companies copied her work, Dids Macdonald launched this not-for-profit initiative in order to raise the profile of intellectual property theft. ACID strives to eliminate the abuse of intellectual property rights by helping its members to understand and protect their ideas. The organisation lobbies to increase the level of damages courts award when plagiarism occurs, and maintains a high profile by attending exhibitions and trade shows across the country and fostering a very close relationship with the media.

ACID has developed several initiatives to encourage a safer trading environment in which to market new inventions. These initiatives include an intellectual property insurance scheme and a range of self-help tools and generic legal agreements for you to use when discussing your invention with manufacturers, advisers, and commercial partners. ACID also offers a mediation service to help its members resolve disputes.

Contact ACID on 0845 644 3617 or e-mail `help@acid.uk.com`. You can also check out the Web site at `www.acid.uk.com`.

Government Support

Business Links provide an invaluable local source of information, advice, and support for people considering how to start or grow a business. From information on how to apply for a government grant to current health and safety regulations, your local Business Link is the best place to start.

Regional schemes operate throughout the UK:

- **Business Link (England):** A government initiative, driven by the Department for Business, Enterprise, and Regulatory Reform (BERR – www.berr.gov.uk). The organisation is made up of a vast network of local Business Links that are connected to one of nine Regional Development Agencies.

  To find out more about Business Link and locate your nearest branch, call 0845 600 9006 or go to www.businesslink.gov.uk.

- **Business Gateway (Scotland):** A Glasgow-based Scottish Enterprise initiative (www.scottish-enterprise.com) that is complemented by 12 Local Enterprise Companies that cover the southern half of Scotland. (Phone: 0845 609 6611; www.bgateway.com.)

- **Highlands and Islands Enterprise (Scotland):** An organisation based in Inverness and managed regionally by nine Local Enterprise Companies that cover Argyll and the Islands, Caithness and Sunderland, Innse Gall, Inverness and East Highland, Lochaber, Moray, Orkney, Shetland and Skye, and Wester Ross. The Highlands and Islands Enterprise is in place to help businesses grow, develop local skills, strengthen communities, and expand global connections. (Phone: 01463 234171; www.hie.co.uk.)

- **Business Eye (Wales):** Welsh entrepreneurs turn to the Business Eye initiative for free independent advice. From employment matters through to finance and taxation, contact Business Eye for the latest information. (Phone: 0845 796 9798; www.businesseye.org.uk.)

- **InvestNI (Northern Ireland):** Another government initiative designed to support innovation, encourage investment, and stimulate entrepreneurial activity within Northern Ireland. Based in Belfast, seven other regional offices represent InvestNI locally. If you're considering starting your own business in Northern Ireland to support the development of your invention, take a look at the InvestNI 'Start a Business Programme' (SABP). Joining is free and you gain access to useful business advice, training, and support. (Phone: 028 9023 9090; www.investni.com.)

British Business Angels Association (BBAA)

Have you ever thought about venturing into *The Dragons' Den*? If so, you're probably aware that preparation is the key when it comes to finding the right business angel to invest in your invention.

The BBAA is a national trade association for the UK's business angel networks. The Department for Business, Enterprise and Regulatory Reform backs the BBAA and requires its members to conform to a code of conduct.

Contact the BBAA (Phone: 020 7089 2305; www.bbaa.org.uk) to find out more about business angels: who they are, what they do, whether they're right for your own venture, what they look for, and perhaps more importantly what *you* must look for.

British Inventors Society (BIS)

The BIS works to improve the national image of independent inventors and the inventing community in general. The BIS was established in 2003 by a group of successful inventors, entrepreneurs, and academics with a common belief that – according to their Web site – 'invention is the vital spark that drives the world's technology and new orders of wealth creation.'

For the latest information on events such as the annual British Invention Show and the British Female Inventor of the Year Awards, visit www.thebis.org. You can register to receive regular e-mail updates and find out where your local inventor club meets. Don't forget to check out the comprehensive links section where you find a vast number of companies, organisations, and support groups to help with the development of your invention. To speak to someone directly, phone 020 8547 2000.

British Library Business and IP Centre

As you may expect, the British Library Business and IP Centre is packed full of useful information to help you turn your bright idea into a successful enterprise.

Visit the Centre to gain free access to online intellectual property and business information databases that would otherwise cost you a lot of money to subscribe to independently. You can swat up on the latest market research reports, business journals, and industry guides to put you ahead of the competition, and you can attend workshops hosted by experienced professionals and successful entrepreneurs.

From individual consultations and personal assistance through to group training, education, and information, the experienced team at the Business and IP Centre offers an invaluable resource to take advantage of.

Simply register for a Reader Pass and use the facilities free of charge! You find the Centre at 96 Euston Road, London NW1 2DB. For more information, call 020 7412 7454, e-mail bipc@bl.uk, or look at www.bl.uk/bipc.

Chartered Institute of Patent Attorneys (CIPA)

Founded in 1882, CIPA is the professional body that represents and examines patent attorneys in the UK. It currently manages over 3,000 members, including trainee patent attorneys and other professionals who are involved with intellectual property matters.

You may consider attending one of the regular inventor clinics that CIPA holds at venues across the UK. You get the opportunity to talk to a registered patent attorney who offers free basic intellectual property advice to assist the development of your idea. Remember that most patent attorneys offer a free initial consultation as well, so if you can't attend a CIPA clinic, you can try contacting a local attorney.

For the benefit of inventors like you, CIPA also has an established code of conduct, which its members must agree to abide by. The code ensures that when you approach a registered patent professional for help, you receive accurate and impartial advice.

If for any reason you think that you received an unacceptable service from a patent attorney (which the internal complaints procedure at the firm itself doesn't resolve) CIPA can assist by investigating your complaint and imposing tough sanctions if it finds genuine professional misconduct.

For more information about the organisation and intellectual property in general, contact the Chartered Institute of Patent Attorneys, 95 Chancery Lane, London WC2A 1DT; phone 020 7405 9450; e-mail mail@cipa.org.uk; Web site www.cipa.org.uk.

ideas21

Enterprise champion Linda Oakley MBE set up the membership organisation ideas21. The organisation gets support from both government and private organisations, including Dyson, the British Library, Business Link, and the UK Intellectual Property Office.

Whether you just have an idea for a new invention or you're looking for a route to market for a finished product, becoming an ideas21 member is a great way to meet other inventors, successful entrepreneurs, and intellectual property professionals.

Membership offers several benefits, including a subscription to regular newsletters, access to evening seminars and networking meetings in London, and free advice sessions for women inventors at the British Library Business and IP Centre.

Think of ideas21 as your access to the 'been there, done that, got the T-shirt' brigade. These experienced and well-connected people can help you avoid the many dangers and pitfalls (and probably some of the costs too) of the invention commercialisation process . . . and in a very friendly manner!

Find out more about the organisation by writing to ideas21, 27 Britton Street, London EC1M 5UD; phone 0208 780 9017; e-mail info@ideas21.co.uk; or visit the Web site www.ideas21.co.uk.

Institute of Trade Mark Attorneys (ITMA)

For over 70 years, the ITMA has been the professional body that represents trade mark attorneys within the UK. It maintains the responsibility for ensuring the specialised knowledge of trade mark professionals and their compliance with the organisation's code of practice.

Find out what a trade mark is, why you may need to register one, and how to go about it by referring to the ITMA Web site at www.itma.org.uk. The site also includes a database that you can search to find a registered trade mark attorney near you.

You can contact the Institute of Trade Mark Attorneys at The ITMA Office, Canterbury House, 2–6 Sydenham Road, Croydon, Surrey CR0 9XE; phone 020 8686 2052; e-mail tm@itma.org; Web site www.itma.org.uk.

Trevor Baylis Brands plc (TBB)

Trevor Baylis, the inventor of the clockwork radio, formed TBB to help the lone inventor overcome many of the problems that he faced throughout the development of his own big idea.

TBB helps inventors just like you evaluate ideas before committing significant time and money to a full-scale marketing effort. The company has set up a sophisticated process of invention assessment through which examiners assess for novelty, market need, and quality.

For ideas that pass the evaluation test, TBB helps you find a route to market by using a network of partners including research organisations, academic institutions, government bodies, and an Industrial Collaborator Network of over 300 manufacturers, distributors, and large retailers – all of which are looking for new ideas to take to market.

For more information contact Trevor Baylis Brands plc, The Enterprise Centre – West Wing, Spelthorne Borough Council Offices, Knowle Green, Staines, Middlesex TW18 1XB; phone 05601 290240; e-mail info@baylis brands.com; Web site www.trevorbaylisbrands.com.

UK Intellectual Property Office (UK-IPO)

For more than 150 years the basic role of the UK-IPO (formerly known as the Patent Office) has remained the same: to promote scientific and technological innovation by securing for inventors the exclusive right to their discoveries for a specific period of time.

The UK-IPO is an official government agency that operates within the Department of Innovation, Universities, and Skills (DIUS) and maintains the responsibility for granting intellectual property rights – most notably patents, registered designs, trade marks, and copyright – to individuals within the UK.

The UK-IPO is also a veritable source of information and can help you find out how to apply for a patent, how to do a patent search on the Web site, how to get information on designs and trade marks, and much more. Contact the UK-IPO at the UK Intellectual Property Office, Concept House, Cardiff Road, Newport, South Wales NP10 8QQ; phone 01633 814000; e-mail enquiries@ipo.gov.uk; Web site www.ipo.gov.uk.

Chapter 24

Ten Inventions (and Inventors) That Changed the World

● ●

In This Chapter

▶ Looking at the inventors whose dreams became reality

▶ Discovering how inventions enhance the quality of our lives

● ●

*M*any of the world-changing inventions came from inventors who made their crucial advances while going about their everyday lives or working with the everyday tasks of their careers. From the caveman's flint tools to the engineer's computer, inventors have made progressive and useful innovations throughout history. When, as here, you look at some of the technological advances of the past three centuries, you can scarcely imagine what inventors'll come up with during the next one. Go to it!

Steaming with James Watt

The famous tale that James Watt invented the steam engine after seeing steam raising the lid of a boiling kettle just isn't true. When he was born in 1736 in Greenock, Scotland, industrial steam engines had been doing real work for years. Most of the credit for the invention of the steam engine must go to Devon-born Thomas Newcomen, who installed the first industrial steam engine in 1712.

Why then do we celebrate Watt and not Newcomen as the father of the steam engine? Well, as with so many inventions, it's the second generation that makes the greatest advances. Watt trained in London as a mathematical instrument maker and then Glasgow University appointed him to this role. In 1765, after working to repair a model of a Newcomen engine for the university, he was wandering through Glasgow Green when he hit upon the idea of making the engine more efficient by using a condenser separate from its steam piston.

His idea resulted in a huge breakthrough. Watt's engine was more powerful and more reliable than Newcomen's and used only about a third as much coal. Watt went on to make further major refinements of the engines: sun-and-planet motion, parallel motion, a smokeless furnace, an air pump, double-action pistons, and a steam jacket for the cylinder.

Watt's radical improvements turned steam into a major power source, and so introduced the age of steam. A stone in Glasgow Green marks the spot where the industrial revolution really began.

By the time of his retirement in 1800 Watt had become a very rich man. In 1882, the British Association for the Advancement of Science gave his name to the unit of electrical power, the *watt*, ensuring his very special place in industrial history.

Cottoning on to Eli Whitney

Eli Whitney is a prime example of both the benefits and hazards of inventing. Born in 1765 in Westborough, Massachusetts, he spent many happy childhood hours tinkering with tools in the workshop where his father made and repaired furniture. After leaving school he gained a place at Yale College, eventually graduating when he was 28.

He found work on a plantation at Mulberry Grove, Georgia, where he made himself useful by inventing devices to help with daily tasks. One of the devices was the cotton gin, a machine that took the seeds out of a bale of the locally grown Green Seed cotton in minutes – in contrast with the days it took workers to hand-pick them. Whitney received a US patent on his cotton gin in 1794, awarded by no less than US founding father, Thomas Jefferson.

Although mainly remembered for the cotton gin, perhaps a more lasting credit for Whitney is that his later inventions in making muskets for the US government used interchangeable parts, which paved the way for the mass production of items, from guns to food grinders. Whitney's mass production innovations made a major contribution to the industrial revolution.

The cautionary part of Whitney's tale comes from the fact that unscrupulous infringers copied the cotton gin and undercut him in price without paying a royalty. Their infringement led to his cotton gin manufacturing company closing its doors in 1797, just three years after Jefferson awarded the US patent. When the patent expired in 1807 the US Congress refused to issue new patents on Whitney's improvements because the gin was so economically valuable.

Writing in 1886, Walter R Houghton, then professor of political science at Indiana University, observed: 'America never presented a more shameful spectacle than was exhibited when the courts of the cotton-growing regions united with the piratical infringers of Whitney's rights in robbing their greatest benefactor . . . [Whitney] had opened the way for the establishment of the vast cotton-spinning interests of his own country and Europe.'

Melting Metal with Sir Henry Bessemer

Henry Bessemer from Hertfordshire accumulated some 110 diverse patents during his lifetime, from diamond polishing wheels to telescopes, graphite pencil leads to solar furnaces. He was a shrewd businessman and innovator who became a millionaire from the manufacture and sales of his various innovations, notably a 'gold' paint made with bronze powder.

During the Crimean War (1853–56) Bessemer invented a more powerful artillery shell but the cast iron gun barrels of the day were too weak to take the extra power and regularly fractured. He realised that steel would make better barrels, but the small quantities available were hugely expensive. So in 1856 he produced the Bessemer Converter, using impure pig iron to make steel in large amounts at lower cost. Essentially, the converter consisted of an egg-shaped vat that held molten iron with all its impurities. Bessemer's invention forced cold air upwards through the vat, floating the impurities to the surface as slag, where they were skimmed off to leave good quality steel below.

The availability of good supplies of Bessemer steel at reasonable prices played a major role in the industrial revolution. Bessemer received a knighthood and was made a Fellow of the Royal Society.

Sparking with Michael Faraday

Michael Faraday, son of a blacksmith, was born in London in 1791. He first trained as a bookbinder but, after listening to lectures at the Royal Institution by its professor of chemistry, Humphry Davy (inventor of among many other things the miners' safety lamp), he applied to Davy for a job.

Davy appointed Faraday to the position of chemical assistant at the Institution, and when Davy retired, Faraday replaced him as professor of chemistry. Faraday's wide chemical research at the Institution covered the discovery of benzene and the condensation of gases, including the first production of liquid chlorine.

Faraday's greatest contribution to science was, however, in the field of electricity. In 1821 he began experimenting with electromagnetism and, by demonstrating the conversion of electrical energy into motive force, he invented the electric motor. In 1831 he discovered the induction of electric currents and made the first dynamo. These advances turned electricity from a scientific curiosity into a practical source of power. Appropriately his name is remembered as the *farad*, the unit of electrical capacitance (the ability of a body to store an electrical charge).

In 1826 Faraday founded the Friday Evening Discourses and the Christmas Lectures for juveniles, establishing his reputation as the outstanding scientific lecturer of the time. Both the Friday Evening Discourses and the Christmas Lectures (televised by the BBC) continue to this day.

Sterilising and Louis Pasteur

Louis Pasteur was born in the eastern French town of Dole, the son of a tanner who was a veteran of Napoleon's armies. At school his teachers didn't regard him as a particularly good student, but he gained good degrees in chemistry and the physical sciences and pursued an academic career at several French universities.

While at the University of Lille, Pasteur received a query from an industrialist on the production of alcohol from beet sugar. It seems that the man's beer was going sour after fermentation. Pasteur correctly suggested that pathogenic micro-organisms already present in the beet were growing during the fermentation. He called these micro-organisms *germs*.

Pasteur believed that he could trace food and beverage spoilage to such pathogens and he studied how they got into the food chain. This led him to the belief that spoilage was preventable by destroying the germs. He correctly concluded that partial sterilisation of a food or beverage, by heating it to about 160°F (80°C) for a period of time, would not greatly change the taste or chemical composition and would preserve useful bacteria while destroying the pathogens and other undesirable bacteria.

The process, appropriately known as *pasteurisation*, remains an important technique in the treatment of milk, beer, and wine, vitally reducing the risks of these liquids causing food poisoning.

Pasteur became a dogged investigator of the scientific basis for all manner of serious diseases. His understanding of fermentation helped him identify the causes of (amongst other diseases) rabies, anthrax, and chicken cholera. He also contributed to developing the first vaccines. In a ranking of the greatest benefactors of humanity, Louis Pasteur would come very near the top.

Staying Alive with Joseph Lister

Essex boy Joseph Lister gained a medical degree in 1852 from University College, London. After a period as an assistant surgeon at Edinburgh Royal

Infirmary, Glasgow University appointed him to the Regius Professorship of Surgery and he later became a surgeon at Glasgow Royal Infirmary.

Hygiene was not a top priority in hospital operating theatres in the middle of the 19th century and Lister noted that there was a very high incidence of post-surgical 'sepsis' infection in wounds. Almost half the patients undergoing surgery died from the subsequent infection.

In a good example of improvements obtained by standing on the shoulders of giants (see Chapter 4), Lister was inspired by reading the work of Louis Pasteur to investigate killing the bacteria that were getting into the wounds from the air. He got the idea that he could destroy the bacteria by applying carbolic acid, found in tar, as an *anti-septic*. So he began to spray carbolic acid onto surgical instruments and on wounds during an operation. He soaked surgical dressings in carbolic acid and applied them to the wounds immediately after the operation.

His work with antiseptics met with initial resistance in the medical community but people soon recognised and widely accepted the effectiveness of his method. By 1860 the surgical mortality rate had reduced to 15 per cent.

Lister's major contribution to public health led to many honours. He served as president of the Royal Society, was named Baron Lister of Lyme Regis, and was one of the 12 original members of the Order of Merit.

Calling Alexander Graham Bell

Born and raised in Edinburgh, Alexander Graham Bell moved with his family to Brantford in Ontario, Canada, at the age of 23 to treat the tuberculosis that threatened his life and had already killed his two brothers. Convalescing in the family's Brantford home he devoted much thought to innovation, and in particular to the field of communications, at least in part as a result of his mother's deafness.

His recovery complete, in 1871 he moved to the United States and settled in Boston, where he began a career of teaching deaf people. His interest in communications was given extra momentum when he fell in love with one of his deaf pupils, Mabel Hubbard, his future wife and the daughter of a wealthy Boston attorney, Gardiner Greene Hubbard.

The concept of the telephone came to Bell on a visit to Brantford in 1874. On his return to Boston he began its development with a young assistant, Thomas Watson. Financial backing came from his future father-in-law and a wealthy leather merchant, Thomas Sanders, whose deaf son was also one of Bell's students. Bell filed his first telephone patent application at the US Patent Office in February 1876 and it emerged on 7 March 1876 as US patent number

174465. On 10 March 1876 intelligible speech was heard for the first time over the telephone when Bell called, 'Mr Watson. Come here. I want to see you.'

News of the telephone quickly spread throughout the US and Europe. By 1878, Bell had set up the first telephone exchange in New Haven, Connecticut. The first telephone exchange in England opened at Coleman Street in London in 1879, serving eight subscribers. Later in that year further exchanges opened in London, Glasgow, Manchester, Liverpool, Sheffield, Edinburgh, Birmingham, and Bristol. Long distance connections followed; the first between Boston, Massachusetts, and New York City in 1884. The telephone had arrived.

Listening to Guglielmo Marconi

The father of radio, Guglielmo Marconi was born in Bologna, Italy, in 1874 and educated in Italy and in England. Back in Bologna in 1894 and with the help of a family friend, Professor Augutus Righi, he conducted his first experiments with his *wireless telegraphy*. These experiments used a crude basic apparatus with multiple oscillators developed by Righi, at first sending signals over a few yards but soon up to a distance of 1.5 miles (2.5 kilometres). Marconi recognised the commercial and military potential and offered the technology to the Italian government who, a bit shortsightedly, turned it down.

The commercial development therefore moved to England where, in March 1896, Marconi filed the first wireless telegraphy patent applications. In 1897 the patent applications emerged as his British patent number 12039. In the same year he opened the first wireless station, at the Royal Needles Hotel on the Isle of Wight. Queen Victoria was duly impressed. With Marconi's help she exchanged wireless messages between Osborne House on the Isle of Wight and the Prince of Wales on board the royal yacht.

The range and uses of wireless telegraphy progressed very fast. In 1899 Marconi transmitted across the English Channel from near Boulogne to Dover. In December 1901, he successfully transmitted a signal from Poldhu, Cornwall, to St John's, Newfoundland, a distance of nearly 2,000 miles (more than 3000 kilometres). People began to establish wireless stations on shore and on military and merchant ships around the world. Marconi's company flourished.

In the popular press his most notable success came in 1910 with the arrest of the infamous criminal, Dr Crippen. Crippen murdered his wife in England and then fled with his new lover to Canada on the *SS Montrose*. The captain, however, became suspicious of the couple and asked his Marconi operator to send a brief message to England. A Scotland Yard detective took a faster ship and arrested Crippen before the *SS Montrose* docked in Montreal.

In 1909 Marconi received the Nobel Prize for Physics, awarded jointly with Professor Karl Ferdinand Braun, a fellow pioneer in wireless telegraphy. After a break serving in the Italian Navy and Army in the First World War, he was much involved with the establishment of public broadcasting in the 1920s. He died in Rome on 20 July 1937. Wireless stations and transmitters all over the world closed down for a period of silence.

Assembling Henry Ford

Henry Ford was born in 1863 in Dearborn, Michigan, into a farming family that was originally from Cork in southern Ireland. From an early age he enjoyed tinkering with machines and he constructed his first steam engine at the age of 15, the year he left school. His engineering skills developed while working in a Detroit machine shop and later as a part-time employee of the Westinghouse Engine Company. He went on to build an internal combustion engine, a single cylinder petrol-fuelled model, leading in 1896 to 'Tin Lizzie' – his first motorcar, with a two-cylinder, four-cycle motor, mounted on bicycle wheels.

He founded the Ford Motor Company in 1903 with the statement: 'I will build a car for the great multitude.' He paid personal attention to his cars' components, as in his US patent number 1005186 for a transmission mechanism. However, just as important as the components of the cars was the way his company put them together. Ford designed an *assembly-line* process in which an unskilled worker learnt one task in the production process and then did that one task on every car. The ability to hire unskilled (and therefore low-wage) workers enabled Ford to sell his cars at a price the average working man was able to afford.

Production started in 1903 with the Model A, followed in 1905 by the Model B, and so on. The major breakthrough came in October 1908 with the introduction of the Model T, for which the opening price was about $900, but over the 19 years of production it fell as low as $280. Ford sold over 15 million in the United States alone.

Ford opened its first factory outside North America at Trafford Park, Manchester, in 1911. The factory produced the Model T. When the factory opened Henry Ford hadn't yet perfected the moving production line and the Trafford Park workmen completed the products on static workbenches. But by 1913 the factory adopted the revolutionary production process in mass production and heralded Britain's first moving production line. Output doubled in the first year and the Model T became the best-selling car in Britain, taking 30 per cent of the market.

Henry Ford died in April 1947. He was a worldwide legend who gave the general public a reliable automobile at an affordable price.

Animating Walt Disney

Walter Elias Disney, known to all as Walt, was born in Chicago on 5 December 1901. He spent much of his childhood in Missouri, where his talent for drawing emerged. Eventually, the family duly moved back to Chicago, where he enrolled at the Chicago Art Institute. He broke off his studies in 1917 to enlist in the US Army for service in the First World War. He spent a year in France, driving a military ambulance that had an unusual feature: a canvas top completely covered by his cartoon drawings.

Back from the war Walt became a commercial artist. He was interested in animation and he and a cartoonist colleague, Ub Iwerks, had early successes with a series of animated short films based on characters from Lewis Carroll's *Alice in Wonderland*. In the 1920s he turned his attention to Hollywood and he persuaded his brother Roy and Iwerks to join him there. He hired an artist called Lillian Bounds to paint the celluloid frames, which was clearly a good partnership because they married in 1925.

With Iwerks he created a cartoon mouse with round ears, initially called Mortimer Mouse. Mortimer's first outings were in silent cartoons called *Plane Crazy* and *Gallopin Gaucho*, which were flops. Walt ordered a new mouse cartoon, *Steamboat Willie*, this time with sound and, with some persuasion from Lillian, Mortimer Mouse became Mickey Mouse. *Steamboat Willie* was a great hit. Walt was the voice of Mickey Mouse and remained so until 1946. Mickey continued to appear in short cartoons and gained some famous friends: Donald Duck, Goofy, and Pluto.

Walt had devised and perfected a camera technique to film through several layers of drawings to create more realistic motion. In 1934 he announced plans to make his first full-length cartoon movie, *Snow White and the Seven Dwarfs*, and in 1937 the first showing received a standing ovation. Other full-length cartoons followed: *Pinocchio, Fantasia, Bambi,* and *Dumbo*. The Second World War interrupted production of some cartoons, including *Alice in Wonderland* and *Peter Pan*, but Walt completed and released these in the late 1940s.

The Disney company rapidly adopted new entertainment avenues, such as television, and branched out into amusement parks, opening the first, Disneyland in California, in 1955. Walt completed plans for Disneyworld near Orlando, Florida, but died in 1966, a few years before it opened.

In the course of his career Walt Disney received 64 nominations for individual Academy Awards and won 26 times, both all-time records. He won the Irving G. Thalberg Award in 1941 for a 'consistently high quality of motion picture production' and in 1963 his *Mary Poppins* won the award for Best Picture.

Chapter 25

Ten Inventors to Emulate

*C*hapter 24 describes ground-breaking inventions of the past, but many of them now feel remote, part of a different age. Although we can gratefully admire the likes of Louis Pasteur, doggedly studying the fermentation of sugar beet, we can't easily identify with him. This chapter turns to people closer to our own time, those who may more easily inspire us to take up our own ideas and bring them to fruition.

Looking at the achievements of inventors such as those we describe in this chapter provides a reminder of how just one person can still make a huge difference and change the world. Many of you have the same talents. Be inspired by the example of these inventors to bring out your own innovations, and develop and apply them – both for your own benefit and for the use, enjoyment and, just sometimes, the adequately expressed thanks of the rest of us.

Sir James Dyson: The Bagless Vacuum Cleaner

James Dyson was born in Cromer, Norfolk. After school, where he showed more interest in Latin and Greek than in science, he went to the Royal College of Art to study furniture and interior design. While at the college, however, he became inspired by the lectures of structural engineer Anthony Hunt and decided to switch to engineering. He gained a thorough grounding in this while working with the engineering firm Rotork, before leaving to start his own company.

The idea for a bagless vacuum cleaner came from his frustration with powdery dust in the old-style vacuum systems clogging their internal tubes and filters. Instead, he decided to experiment with an industrial cyclone tower, a system known for removing suspended dust from air streams. Turning this concept into a reliable user-friendly domestic cleaner took around five years and over 5,000 prototypes. He offered his invention to major manufacturers, but they all turned it down.

Eventually, James Dyson found a Japanese licensee. Manufacture and marketing of the first production model (the G-Force) started in 1986. It won the 1991 International Design Fair prize. The Japanese licensing income funded the development of versions to collect finer particles of dust. The result was the DC01, the first in a range giving constant suction. The DC01 became the best-selling vacuum cleaner ever.

James Dyson has won numerous awards for his work, including the Price Waterhouse West of England Business of the Year and the Prince Philip Designers Prize in 1997, and the Designer of the Decade in 1999. He received a CBE in 1998 and was knighted in the New Year's Honours list for 2007.

Georges de Mestral: The Velcro® Hook-and-Loop Fastener

Georges de Mestral (1907–90) was an inventive Swiss engineer, born and raised near Lausanne. He received his first patent – for a model aircraft – when he was just 12. By taking casual work he paid for his engineering studies at a Lausanne polytechnic and then he took a job with a Swiss engineering company, which still gave him time to pursue his inventing.

On holiday one year hunting game in the Jura Mountains he was plagued by burrs from burdock plants clinging to his clothes and to his dog's coat. Later he examined the burrs under a microscope and saw that their barbs hooked into the looped fibres of his clothes. He realised that he could bind two materials together in a similar way. He enlisted the help of a textile weaver and a loom maker and created his 'hook-and-loop' fastener. Development took years and early trial versions suffered from the arguable drawback that skirts and bras that were held together with the new fastener tended to fall off.

Georges de Mestral filed his first patent application on the fastener in Switzerland in 1951, which emerged in 1954 as Swiss patent number 295638. He coined the name *Velcro*® for the fastener, from the French 'velours' (velvet) and 'crochet' (hook). First registered as a trade mark in 1958, the term 'Velcro' is now so widely used for any hook-and-loop fastener that it risks becoming 'generic', which would invalidate the registrations (see Chapter 8 for details of how that can happen). Velcro International put much effort into reminding the public that Velcro is its registered trade mark.

Dr Alex Moulton: Rubber Suspension for the Mini® and the Moulton® Bicycle

Alex Moulton was born into a Wiltshire family whose business in the rubber industry began in 1848 with Stephen Moulton, his great grandfather. Stephen Moulton himself deserves a mention as the person who brought American rubber manufacturing expertise to the UK and invented crucial improvements in vulcanising rubber to increase its strength.

In 1956 the family firm became part of the Avon Rubber Company and Dr Moulton founded his own company, Moulton Developments Limited, working closely with the then British Motor Corporation (BMC) on car suspension systems. These systems included the conical rubber springs that, when used with small wheels, allowed the innovative design of Sir Alec Issigonis for the BMC mini to become a reality. Further developments included the 'hydrolastic' suspension for the Austin 1100/1300, the UK best-selling car of its day, and 'hydrogas' suspension used for the Rover Metro.

As well as developing car suspensions, Dr Moulton created his revolutionary Moulton® bicycle, first marketed in 1962. Again, it employed small wheels and a rubber suspension, making it the original full-suspension bicycle. He has continued to develop the bicycle over many years, producing a wide range of models and a massive following of devotees in many countries – notably in Japan, where the bike has a cult status.

Winner of many design awards, Dr Moulton is a Fellow of the Royal Academy of Engineering and served as its vice president from 1985–88. He was awarded the CBE in 1976 for services to industry.

James L. Fergason: The Liquid Crystal Display (LCD)

James Fergason was born in Wakenda, Missouri, and gained a degree in physics at the University of Missouri. Moving on to a research position at the Westinghouse Research Labs in Pennsylvania, he set up the first American research team to study liquid crystals. Later, he became associate director of Kent State University's Liquid Crystal Institute.

Friedrich Reinitzer of Austria and Otto Lehmann of Germany had discovered liquid crystals in the 1880s, but development of the practical applications did not start until the late 1950s. Fergason's first patent for a liquid crystal display was US patent number 3410999. The 1971 breakthrough for a 'nematic liquid crystal twist cell display' came with his US patent number 3627408. He founded a company to manufacture the displays and sold them to a Swiss watchmaker.

Since then, James Fergason has built up a huge array of over 600 patents on liquid crystals (more than 125 in the United States alone) that support the huge LCD industry created when he demonstrated the first practical versions in 1971. By 1977 LCDs had largely replaced light emitting diodes (LEDs) in digital watches: they were cooler and used much less energy. Since then LCDs have remade nearly every type of information display on consumer electronic devices and on industrial, scientific, and medical apparatus.

James Fergason has won many awards for his work and in 1998 he was inducted into the US National Inventors Hall of Fame.

Mandy Haberman: The Haberman® Feeder and the Anywayup® Cup

Hertfordshire-based Mandy Haberman, a graduate of London's St Martins School of Art, was drawn into becoming an inventor by a need for an improved feeder for a baby daughter, who had difficulty with sucking. Finding no suitable feeding bottle on the market she developed her own, the Haberman® feeder, and in 1987 she set up a mail order company to supply it. Now made and sold through licensees, hospitals and parents around the world use the feeder to assist babies with sucking problems. Tracy Hogg in 'The Baby Whisperer' has identified it as 'an excellent product for transitioning babies from breast to bottle.'

Some years later Mandy was inspired to invent a leak-proof cup for young children, after seeing a toddler spill blackcurrant juice onto a cream-coloured carpet. The result was the Anywayup® trainer cup, which has a spout with a slit valve to control the liquid flow. Her first patent application on the cup, filed in 1992, became GB patent number 2266045. Overseas patents followed. Although several makers of infant products showed initial interest in taking a licence, no licensing agreement emerged. So she formed her own company in 1996 and teamed up with a marketing company in South Wales to launch the product. By the next year annual sales grew to £750,000, helped by much word-of-mouth recommendation from mothers of small children.

Mandy went through an experience that many start-up businesses fear. Just a few months after the launch of the Anywayup® cup she found that one of the possible UK licensees she'd approached was now offering a very similar product. She sued them, using her house as security to support the legal costs. The action went her way. The court found that her patent was valid, that the rival product infringed it, and ordered an injunction against further infringement. The company appealed but abandoned the appeal after reaching an out of court settlement in Mandy's favour.

Mandy has won many awards, including the British Female Inventor of the Year in 2000 and a Gold Medal from the Salon International des Inventions in Geneva. The Anywayup® cup has won awards too, including the DBA Design Effectiveness Award for innovation and product design and the British Plastics Federation Award for innovative use of plastics.

Ron Hickman: The Workmate® Work Bench

Born and educated in the Natal province of South Africa, Ron Hickman moved to England in 1955. He became engineering director of the Lotus car company, working with the designer Colin Chapman. Together, they created the legendary Lotus® Elan® and the Lotus® Europa™. For the Europa they created a design giving a level of air resistance to the car's motion that was remarkably low for its day and is still difficult to achieve.

One weekend in 1961, while indulging in do-it-yourself household duties, Ron Hickman sawed through a dining room chair he was using to support his work. He got the feeling that a purpose built support may have been more sensible and began to think about what form it could take. Several years passed before he perfected a prototype, but by 1968 he'd filed his first patent application (granted as GB1267032).

With the prototype ready, he approached Black & Decker to offer a manufacturing licence. The company turned it down, as did several other organisations. So he began to manufacture the work benches himself, selling them by mail order. Four years and 14,000 unit sales later Black & Decker revived negotiations and the two parties agreed a licensing deal. The Hickman granted patents proved vital as infringers appeared around the world, notably in the US and Japan, but in every case the infringement actions succeeded.

Mass production of the Workmate® workbench began in 1973. Hickman sold 30 million units worldwide while the original UK and equivalent patents were in force, then global sales continued to a total in excess of 65 million units.

The Workmate® royalties have made Hickman a wealthy man and he retired to Jersey to continue his inventing activities. In 1977 the Queen awarded him with an OBE for industrial innovation.

Stephanie Kwolek: Kevlar® Ultra-High-Strength Fibre

Stephanie Kwolek was born in New Kensington, Pennsylvania. As a child she was influenced by her father's interest in the plants and seeds they discovered when roaming together in the woods near their home. Her father died when she was ten but his spirit of curiosity never left her.

She studied chemistry at the University of Pittsburgh, then took a post with du Pont's textile fibres laboratory in Buffalo, New York State. du Pont's William H Carothers had created nylon, the first wholly synthetic fibre, in the early 1930s and Stephanie joined a team investigating new synthetic fibres and how to make them.

Her research led to a new group of polymers with long, straight, rigid molecules, rather like very tiny matchsticks (see her US patent number 3819587). She discovered that when these polymers dissolved they formed liquid crystals, which no polymer had ever done before. When spun, the molecules lined up to form a fibre that was exceptionally strong. du Pont realised her fibre had huge potential and, after further work and refinement, the company launched it in 1971 under the trade mark Kevlar®.

Kevlar® fibre is heat- and flame-resistant, five times stronger than steel, and lighter than glass fibre. Manufacturers use it in hundreds of products, from tyres and brake pads, fibre-optic cables, skis, and surfboards, to composites for spacecraft and cables for bridges. The most famous application for Kevlar® is in bullet-proof vests, and in these Kwolek's invention has saved thousands of lives.

Stephanie Kwolek's many awards include the Kilby Award and the US National Medal of Technology. In 1999, in recognition of her own pioneering work and her encouragement of the next generation of innovators, she won the Lemelson-MIT Lifetime Achievement Award.

Owen Maclaren: The Maclaren® Baby Buggy

Owen Maclaren was born in Saffron Walden, Essex, a descendant of the Maclaren clan from Argyllshire. He was an inventive engineer who also served as an aircraft test pilot. His work before the Second World War included the design of the undercarriage for the Spitfire fighter aircraft, and such design work gave him a good knowledge of strong, rigid structures that would fold neatly.

The baby buggy story began in 1965 when Maclaren's daughter and his first grandchild came to visit. He found himself faced with the task of wheeling the child around in a heavy conventional pushchair. Clearly, he could do better than that. So in the stables of his farmhouse home in Northamptonshire he started to develop a lightweight pushchair that safely carried a child but that folded into a small space after use. He built his prototype from aluminium – it weighed just 6 pounds (2.75 kilograms) and folded to a size not too much bigger than a rolled-up umbrella. He filed a British patent application on 20 July 1965, leading in 1968 to patent GB1154362 and the equivalent US patent number 3390893.

The buggy proved a great hit with the public. By 1976 Maclaren's annual output was nearly 600,000 units, consuming nearly 4,000 miles (6,400 kilometres) of aluminium tubing and 1.5 million yards (1.4 million metres) of fabric.

Owen Maclaren was awarded an MBE in the 1978 New Year Honour's List, but died later that year at the age of 71.

Shigeru Miyamoto: Donkey Kong® and the Nintendo® Games Empire

Shigeru 'Shiggy' Miyamoto was born and grew up in near Kyoto, Japan. From an early age he had a wild imagination and a great sense of adventure, helped by exploring the fields, streams, hills, and caves near his home. He entered the Kanazawa Munici College of Industrial Arts and Crafts to study industrial design, but took five years to graduate and still wasn't able to reach a decision on a career. He had thought of becoming a painter or a puppeteer, and later made toys as means of keeping occupied.

In 1977, aged 24, Miyamoto decided to approach the head of toy company Nintendo, Hiroshi Yamauchi, who helpfully was an old friend of his father. At the meeting Yamauchi asked him to come up with some ideas for new toys. Miyamoto came back with a bagful of new ideas. Duly impressed, Yamauchi appointed him to the new position of company staff artist.

In 1980 Yamauchi decided that he wanted a new video game and asked Miyamoto to create one. Miyamoto's imagination and artistic skills produced Donkey Kong®, which marked the beginning of Nintendo's huge successes in video games. Super Mario Bros® and The Legend of Zelda soon followed.

Miyamoto and his teams have built Nintendo into a games powerhouse. The games appeal mainly, but by no means exclusively, to children, because players can readily understand and control the simple characters' adventures. Sales of the games and the systems to play them have generated revenues of more than £5 billion. Miyamoto also headed up the team developing Wii® systems and games. In the nine months from their launch in November 2006, Nintendo sold some 9 million Wii units around the world.

Miyamoto is one of the most successful modern artists and his peers regard him as the greatest game designer ever. He's received many awards, including the Game Developers Choice – Lifetime Achievement Award in 2007.

Dr Forrest M. Bird: Respirators

Forrest Bird's respirators help patients in about every hospital in the world. He developed the first small mass-produced respirator in the 1950s (the Bird Mark 7), which eliminated the iron lung (a huge tank slightly larger than the patient that the patient actually lived inside). A keen and experienced aviator, he tested the Bird Mark 7 device by travelling in his own aircraft to medical schools and asking doctors for their most seriously ill patients. In each case, known therapies had failed and doctors expected the patient to die of heart failure. The respirator couldn't save every patient, but even the failures pointed the way for further improvements in the device.

In 1969, Dr Bird developed the Babybird: the first effective respirator for very small premature babies. Before his invention, over 70 per cent of premature babies with major heart and lung problems died. In two years, the Babybird reduced this death rate to below 10 per cent.

Dr Bird built his company, Bird Corporation, through his own efforts with no stockholders and no investors. The corporation manufactured his many medical respirator designs and operated completely debt-free. In due course Bird Corporation merged into the 3M corporation.

Although he gets this last place in the last chapter of this book, Dr Bird is anything but the least of those deserving a special mention in it. The husband of Dr Pamela Riddle Bird, the author of the original *Inventing For Dummies* on which we based this UK edition, he was an inspiration and key supporter of her work in researching and writing the book. He wrote the foreword to get that original off to a flying start.

We have earlier expressed our deep and grateful thanks to Dr Pamela Riddle Bird for the basis and structure of this edition, but it does seem entirely right that our thanks to Dr Forrest Bird provide its last word.

Part VII
Appendixes

'OK – I've put on my safety helmet made
from my own impact - proof material,
now where's the testing area?'

In this part . . .

They may look a little dry, but the sample forms that make up the first Appendix to this book could save you a ton of heartache. You'll also find a useful listing of online resources for the budding inventor.

Appendix A

Sample Agreements

. . . .

*H*ere we provide you with some sample agreement forms to cover the nitty-gritty of confidentiality, work-for-hire and product licensing.

SAMPLE CONFIDENTIALITY AGREEMENT

AGREEMENT and acknowledgement between _____ (Owner) and
_____ (Recipient).

Whereas, the Owner agrees to furnish the Recipient with certain confidential information relating to his or her business affairs for purposes of:

Reviewing, Analyzing and Evaluating for possible purchase and/or license the rights to and/or for manufacturing and/or commercial development of the Owner's
_____ ;

Whereas the Recipient agrees to review, examine, inspect or obtain such information only for the purpose described above, and to otherwise hold such information confidential pursuant to the terms of this Agreement,

BE IT KNOWN, that the Owner has or shall furnish to the Recipient certain confidential information (including, but not limited to the items on the attached list), and may further allow the Recipient the right to interview the Owner, all on the following conditions:

1. The Recipient agrees to hold all confidential or proprietary information or trade secrets ("information") in trust and confidence and agrees that the information shall be used only for the contemplated purpose, and shall not be used for any other purpose or disclosed to any third party.

2. At the conclusion of our discussions, or upon earlier demand by the Owner, all information, including written notes, photographs, memoranda, photocopies of the confidential material or notes taken by you, the Recipient, shall be returned to the Owner.

3. This information shall not be disclosed to any employee or consultant unless they agree to execute and be bound by the terms of this Agreement.

4. It is understood that the Recipient shall have no obligation with respect to any information known by the Recipient prior to the date of this Agreement (other than information disclosed to Recipient by Owner) or generally known within the industry prior to the date of this Agreement, or which becomes common knowledge within the industry thereafter through means other than by default of this Agreement by Recipient, or after ten (10) years from the date of this agreement.

5. This Agreement may be modified only by a writing, signed by both parties.

(Owner)

Dated: _____   By _____

(Recipient)

Dated: _____   _____

Signed By_____

Title_____
For _____

SAMPLE CONSULTANT OR EMPLOYMENT AGREEMENT:
INNOVATION AND PROPRIETARY INFORMATION

In consideration of my engagement as an independent consultant or employee by
_____ (the Company), I agree to:

1. Disclose and assign to the Company, as its exclusive property, all inventions, technical or business innovations developed or conceived by me solely or jointly with others during the period of my engagement, (a) that are along the lines of the businesses, work or investigations of the Company or its affiliates to which my engagement relates or as a result of which I may receive information due to my engagement, or (b) that result from or are suggested by any work which I do for the Company, or (c) that are otherwise made through the use of Company time, facilities or materials;

2. Make and maintain for the Company adequate and current written records of all such inventions or innovations;

3. Execute all necessary papers and otherwise provide proper assistance (at the Company's expense) during and subsequent to my engagement, to enable the Company to obtain for itself or its nominees or agents, any patents, or other legal protection for such inventions or innovations in the United Kingdom and abroad; and

4. When my engagement terminates, to promptly deliver to the Company all written and other materials which are of a secret or confidential nature relating to the business of the Company or its affiliates. The terms "secret" and "confidential" are used in the ordinary sense and include materials, information and data such as drawings, manuals, notebooks, reports, models, inventions, formulas, processes, machines, compositions, computer programs, accounting methods, business plans and information systems;

5. Not to use, publish or otherwise disclose (except as required by the Company), either during or subsequent to my engagement, any secret or confidential information or data of the Company or any information or data of others which the Company is obligated to maintain in confidence; and

6. Not to disclose or utilise in my work with the Company any secret or confidential information of others (including any prior employers), or any inventions or innovations of my own which are not included within the scope of this Agreement.

7. The obligation not to disclose secret or confidential information shall remain in effect for ten (10) years after my engagement terminates or when the information becomes generally known in the industry.

8. This Agreement supersedes and replaces any existing agreement between the Company and me relating generally to the same subject matter. It may not be modified or terminated, in whole or part, except in writing signed by an authorised representative of the Company. Discharge of my responsibilities in this Agreement is an obligation of my executors, administrators, or other legal representatives or assigns.

_____ (Signed/Dated)   _____ (Signed/Dated)
For the Company   Consultant or Employee
(Authorised signatory)

_____   _____
Witness   Date   Witness   Date

SAMPLE ASSIGNMENT AND
SERVICE CONTRACT

AUTHOR: _____

_____
(List full name and address of author.)

EMPLOYER: _____

_____
(List full name and address of employer)

      WHEREAS the Employer desires to hire and employ the above referenced Author to write, draft and/or edit instructional text and materials (including any accompanying illustrations, designs, diagrams or other graphic representations); and

      WHEREAS the Author desires to perform services, it is hereby agreed by and among the parties hereto as follows:

<u>1. DESCRIPTION OF WORK:</u> The Employer hereby retains the services of the Author to perform services as an author, writer and/or editor for the creation of instructional text and materials (including any accompanying illustrations, designs, diagrams or other graphic representations), whether in printed or electronic form, to be used in connection with the Employer's instructional text and materials.

<u>2. ASSIGNMENT OF COPYRIGHT:</u> The Author acknowledges and agrees that his/her services are prepared within the scope of this Agreement as services undertaken for the Employer. The Author further acknowledges that Employer is the exclusive owner of all copyright with respect to the instructional materials (including any accompanying illustrations, designs, diagrams or other graphic representations) actually used by the Employer and all preliminary drafts leading to the final draft actually used. The Employer has the right to exercise all rights of the copyright proprietor with respect thereto. In the event that the Author is ever deemed to be the copyright owner in the instructional materials subject to this Agreement, the Author hereby assigns, transfers, sets over and conveys to the Employer and his successors and assigns that portion of all right, title and interest set forth above in and to the printed instructional materials including the copyright and proprietary rights therein and in any and all versions or derivatives regardless of the media used whether physical or electronic and now known or hereinafter developed, and any renewals and extensions thereof (whether presently available or subsequently available as the result of intervening legislation) in the United Kingdom and elsewhere throughout the world, and further including any and all causes of action for infringement of the same, past, present and future, and all proceeds from the foregoing accrued and unpaid and hereafter accruing. The Author agrees to sign any and all other papers which may be required to effectuate the purpose and intent of this Assignment, and hereby irrevocably authorises and appoints the Employer as the Author's authorised representative to take such actions and make, sign, execute, acknowledge and deliver all such documents as may from time to time be necessary to convey to the Employer or his successors and assigns, all rights granted herein.

3. <u>FURTHER WORKS:</u> The Author further agrees that if the Employer should ask him/her to draft new, revised, updated or additional instructional materials, and if the Author agrees to do so, any and all such materials will also be deemed to be a work made for hire, and that in the event a determination is ever made that such materials do not constitute work made for hire, the transaction will be deemed to include and provide for the transfer to the Employer all copyright, Author's rights and other intellectual property rights in and to such new, revised, updated or additional instructional materials, as set forth in paragraph 2 above.

4. COLLABORATION: The Author will not use assistants or other writers in collaboration with creating the instructional materials subject to this Agreement without express approval from the Employer and any such person or persons, if approved of by the Employer, must expressly agree in writing to be bound by the terms of this Agreement.

5. ENTIRE AGREEMENT: This Assignment/Work for Hire Agreement sets forth the entire agreement between the parties with respect to the subject matter hereof, and no modification, amendment, waiver, termination or discharge of this Agreement, or any provisions hereof, shall be binding upon any party unless confirmed by written instrument signed by all parties.

6. NON-WAIVER: No waiver of any provisions or default under this Agreement shall affect the right of any party thereafter to enforce such provision or to exercise any right or remedy in the event of any other default, whether or not similar.

7. SEVERABILITY: If any part of this Agreement is determined to be void, invalid, inoperative or unenforceable by a court of competent jurisdiction or by any other legally constituted body having jurisdiction to make such determination, such decision shall not affect any other provisions hereof, and the remainder of this Agreement shall be effective as though such void, invalid, inoperative or unenforceable provision had not been contained herein.

8. GOVERNING LAW: The validity, construction and effect of this Agreement and any and all extensions and/or modifications thereof shall be governed by English law. In any legal action between the parties, the prevailing party shall be entitled to recover damages, injunctive or other equitable remedies, its court costs and reasonable attorney's fees. This Agreement shall be binding upon the parties hereto, their heirs, personal representatives, successors and assigns but shall not take effect until executed by the Author and the Employer.

   IN WITNESS WHEREOF, the undersigned have executed the foregoing Agreement as of this _____ day of _____, 20__.

AUTHOR:

_____

EMPLOYER:

By:_____

SAMPLE MANUFACTURING CONTRACT

This sample contract is provided mainly for the purpose of illustration. It includes certain features, for example specimen figures for prices, percentages and time periods, that require adjustment for each individual contract

(Your Letterhead)

(Date)

TO:

Subject: Contract Manufacturing Agreement

Dear Mr _____ :

We are pleased to submit this proposal for (The Manufacturer's Name), hereinafter called the "Manufacturer," to perform contract manufacturing services to (Your Company's Name), hereinafter called the "Company," for our (Your Product), hereinafter called the "Product."

The Parties herein agree that:

The Company shall purchase from Manufacturer and Manufacturer shall exclusively supply to the Company items and products listed in Exhibit 1 attached. The parties further agree that only Company's orders shall be valid and binding when written on Company's purchase orders and transmitted from the Company to the Manufacturer.

The Company may, at its option, cancel this Agreement before the stated termination date upon the occurrence of:

1. Presentation of a true and verified proposal from a competitive manufacturer of a unit price of 5% or greater, LESS THAN THE PRICE stipulated in this Agreement. However, the Manufacturer has an option to modify the cost by reducing the unit cost to within 3% of the verified competitive proposal and thereby render this Agreement NOT subject to cancellation under this provision.

2. Delays of 30 days or more in production from the date of the Company's postmarked purchase order to the Manufacturer's shipment date unless waived in writing from the Company.

3. Manufacturer's declaration of bankruptcy or insolvency.

4. Acquisition of Manufacturer by another company or entity.

5. Majority of stock issued by Manufacturer being purchased or acquired by another firm or person.

Payment terms are two percent (2%) ten days, net thirty (30) days. The Manufacturer is not required to continue shipping if the Company does not meet its payment terms.

The Parties hereto further agree that all products, drawings and samples, whether complete or in progress, are the property of the Company. The Manufacturer agrees to assign all patents and products improvements to the Company. In the event of termination of this Agreement, if the Company changes its source of supply as above listed, the Manufacturer agrees not to manufacture Products listed in Exhibit 1 for any other firm or person, for a period of no less than (5) years beginning from the termination date.

The Manufacturer agrees that the Company owns all moulds, dies and fixtures used in the manufacture of the Company's Products, except for those that the Company has not paid or has not been invoiced. In the event the Company elects to change its source of supply, as stated above, the Manufacturer either will immediately deliver (delivery carrier to be specified by the company) to the Company all such tools, dies, molds and fixtures or will sell all said tools, dies, molds and fixtures at their current book values as calculated by generally accepted accounting standards, at the Company's option. (The proceeds of said sale will also immediately be delivered to said Company). The Company shall be entitled to seek immediate injunctive relief to obtain possession of these items.

The Manufacturer also agrees that no employee will use knowledge of any of the Products gained during their employment with the Manufacturer to manufacture or market the Company's Products or similar products either directly or indirectly or to advise, counsel or in any way assist any third party to compete with the Company. Furthermore the Manufacturer agrees not to use any information provided by the Company, directly or indirectly, for the Manufacturer's own benefit or for the benefit of any other person, firm or corporation.

The Manufacturer agrees to use its best efforts to insure high quality products and workmanship. The Manufacturer warrants all products sold to the Company under this Agreement will be free from defects due to poor workmanship or material for a period of thirteen (13) months. If Products are defective the Manufacturer will be responsible to replace the defective Products without charge, credit the Company's account, or issue the Company a refund, at the Company's option. The warranty shall supersede all previous warranties of any kind whether written or implied.

The Company agrees to keep the Manufacturer apprised of product sales; and the Company further agrees to use its best efforts to sell the Products.

Either Party may terminate this Agreement by giving ninety (90) days written notice sent via Recorded Delivery mail. If this Agreement is terminated, the Company has the sole option to purchase the Manufacturer's inventory of said Products over a six (6) months period. If the Company elects not to purchase the Manufacturer's inventory of Products, the Manufacturer may dispose of such finished goods inventory in any method it chooses.

The terms of this Agreement shall survive in the event of termination.

This Agreement is not assignable by either Party except by written agreement by both Parties. If this Agreement is assigned, the assignee shall be bound by the terms and conditions of this Agreement.

Correspondence shall be addressed to the individuals on Page 1 of this Agreement.

Exhibit 1 - Products

Product(s) to be manufactured or supplied by the Manufacturer:

1.
2.
3.

Agreed By:

_____   _____
(Manufacturer)   (Company)

_____   _____
(Signature of officer)   (Signature of officer)

_____   _____
(Print name and title)   (Print name and title)

_____   _____
(Date)   (Date)

SAMPLE FRAMEWORK FOR A LICENCE AGREEMENT

This sample framework indicates features to be covered in a typical licensing agreement. The fine detail needs to be added, or its presence confirmed, for each individual licence.

THIS AGREEMENT is made this .... day of .... (month) .... (year) at ....(place)

BETWEEN

1.1 .... AND
Name and address of the Licensor (the person or company offering the licence).

1.2 ....
Name and address of the Licensee (the person or company receiving the licence).

WHEREAS the Licensor is the owner of certain proprietary rights in regard to

1.3 ....
(general definition of the matter to be licensed)
including granted patents*/patent applications*/design rights*/trade marks* as defined in the attached Schedule, together with associated know-how*
[*include as required, and list the individual rights in an accompanying Schedule].

AND the Licensee wishes to acquire those rights according to the terms of this agreement

NOW IT IS HEREBY AGREED:

2. Definition of terms used in the agreement.

3. Scope of Licence:
   3.1 Exclusive*/Sole*/Non-exclusive* [*select as required]
       3.1.1 Rights to sub-license
       3.1.2 Requirement for Licensor to consent to each sub-license
       3.1.3 Liability for actions of Sub-licensee.
   3.2 Manufacture*/Use*/Sale*/Lease* [*include as required].
   3.3 List of territories included.
   3.4 Warranties from Licensor or Licensee.
   3.5 Inclusion or exclusion of improvements by the Licensor or Licensee
   3.6 Responsibility for securing and maintaining patent and other IP rights
   3.7 Ability to take over 3.6 actions if the party responsible fails to act.

4. Provision of technical assistance and information by Licensor
   4.1 Confidentiality requirements from Licensee and Sub-licensees.

5. Payments
   5.1 Lump sum payment(s)
   5.2 Royalty rates
       5.2.1 Flat rate, or
       5.2.2 Rate bands according to number of units

5.3 Minimum royalty or minimum unit numbers
5.4 Requirements and timing for royalty or unit numbers
    5.4.1 Nature and frequency of reporting royalty or unit numbers
    5.4.2 Licensor's right to inspect Licensee/Sub-licensee records
    5.4.3 Payment schedule for royalties
    5.4.4 Penalties if payment schedule not met.

6.   Infringement by a Third Party
6.1 Licensor or Licensee primarily responsible for taking action against infringement.
6.2 Licensee or Licensor's rights to take part in infringement action initiated by the other.
6.3 Apportionment of costs incurred in taking infringement action
6.4 Apportionment of damages and any other remedies received from infringement action.

7.   Infringement by Licensee
7.1 Liability of Licensor for infringement by Licensee
7.2 Licensor's right to take part in Licensee's defence to infringement action.
7.3 Apportionment of costs incurred in defensive infringement action.

8.   Period of agreement
8.1 Defined time period in years, or
8.2 Set by expiry of defined patent
8.3 Provisions for early termination by Licensor or Licensee
8.4 Rights and obligations of the Licensor and Licensee after termination.

9.   Effect of change of status of the Licensor or Licensee
9 1 Provisions in the event of winding-up of Licensor or Licensee
9.2 Provisions in the event of takeover of Licensor or Licensee.

10.   Applicable law (e.g. English).

11.   Licensor and Licensee's Addresses for Service.

12.   Legal status of signatories for Licensor and Licensee.

SIGNED, the day, month and year written above

For and on behalf of the Licensor   For and on behalf of the Licensee

………………………………………   ………………………………………..
(Name)   (Name)

………………………………………   ………………………………………..
(Position)   (Position)

Appendix B

Sources of Help

● ●

*H*ere we list Web site details for a variety of organisations that can help you protect, develop, and market your ideas.

The lists are not exhaustive. The details given are believed to be correct at the date of publication but the authors can guarantee neither their accuracy nor the quality of service the organisations may provide for you. No payment has been sought or received by the authors from any of those named.

Intellectual Property Offices

In this section, we list intellectual property law contacts on an international basis. With global markets expanding, this list enables you to investigate your invention's patentability around the world.

- ✔ African Regional Intellectual Property Office (ARIPO): www.aripo.org
- ✔ Eurasian Patent Organisation (EAPO): www.eapo.org
- ✔ European Patent Office (EPO): www.epo.org
- ✔ EU Plant Variety Office (CPVO): www.cpvo.europa.eu
- ✔ International Union for the Protection of New Varieties of Plants (UPOV): www.upov.int
- ✔ Internet Corporation for Assigned Names and Numbers (ICANN): www.icann.org
- ✔ Irish Patent Office: www.patentsoffice.ie
- ✔ Japanese Patent Office: www.jpo-miti.go.jp
- ✔ Office for Harmonisation in the Internal Market (OHIM): oami.europa.eu
- ✔ Organisation Africaine de la Propriété Industrielle (OAPI): www.oapi.wipo.net
- ✔ United Kingdom Intellectual Property Office (UK-IPO): www.ipo.gov.uk
- ✔ United States Patent & Trademark Office (USPTO): www.uspto.gov

- World Intellectual Property Office (WIPO): www.wipo.int
- UK Plant Variety Rights Office (PVRO): www.defra.gov.uk/planth/pvs

Governmental Departments and Offices

Check out these sites for important legislative and regulatory information.

- Companies Registry (Northern Ireland): www.detini.gov.uk
- HM Revenue & Customs: www.hmrc.gov.uk
- Manufacturing Advisory Service: www.mas.dti.gov.uk
- Office for National Statistics: www.statistics.gov.uk
- UK Companies House: www.companieshouse.gov.uk
- UK Department for Business, Enterprise & Regulatory Reform(BERR) [formerly part of the DTI]: www.berr.gov.uk
- UK Department for Environment, Food and Rural Affairs (DEFRA): www.defra.gov.uk
- UK Trade & Investment: www.uktradeinvest.gov.uk
- United Nations Educational, Scientific & Cultural Organisation (UNESCO): www.unesco.org
- World Trade Organisation (WTO): www.wto.org

Professional Bodies

Chances are, you'll need legal and professional advice in taking forward your career as an inventor. These sites can help.

- American Bar Association: www.abanet.org
- International Association for the Protection of Intellectual Property (AIPPI): www.aippi.org
- British Business Angels Association: www.bbaa.org.uk
- Chartered Institute of Patent Attorneys (CIPA): www.cipa.org.uk
- Chartered Society of Designers: www.csd.org.uk
- Confederation of British Industry: www.cbi.org.uk
- European Patent Institute (EPI): www.patentepi.com
- The Institute of Directors: www.iod.com
- The Institute of Trade Mark Attorneys (ITMA): www.itma.org.uk

- The Institution of Electrical Engineers: www.theiet.org
- The Institution of Mechanical Engineers: www.imeche.org
- The Law Society: www.lawsociety.org.uk
- Licensing Executives Society [Europe]: www.les-europe.org
- Licensing Executives Society [USA & Canada]: www.usa-canada.les.org
- Regional Development Agencies (England): www.englandsrdas.com:
- The Royal Academy of Engineering: www.raeng.org.uk
- The Royal Institution: www.rigb.org
- The Royal Society: www.royalsoc.ac.uk
- Union of European Practitioners in Intellectual Property (UNION): www.union-eu.com

Support Organisations and Businesses

Inventing can be a lonely furrow to plough. These links offer inspiration, support, and advice.

- Alliance against IP Theft: www.allianceagainstiptheft.co.uk
- Anti Copying in Designs (ACID): www.acid.uk.com
- The British Association of Women Entrepreneurs: www.bawe-uk.org
- The British Inventors Society: www.thebis.org
- The British Library: www.bl.uk
- The British Venture Capital Association: www.bvca.co.uk
- Business Eye (Wales): www.businesseye.org.uk
- Business Gateway (Scotland): www.bgateway.com
- Business Link: www.businesslink.gov.uk
- The Carbon Trust: www.carbontrust.co.uk
- The Design Council: www.designcouncil.org.uk
- The Engineering Council: www.engc.org.uk
- Federation Against Copyright Theft (FACT): www.fact-uk.org.uk
- Federation Against Software Theft (FAST): www.fast.org.uk
- Highlands and Islands Enterprise (Scotland): www.hiebusiness.co.uk
- Ideas21 Limited: www.ideas21.co.uk
- InvestNI (Northern Ireland): www.investni.com

- ✔ Official James Dyson site: www.dyson.com/about/story
- ✔ Official Mandy Haberman site: www.mandyhaberman.com
- ✔ The Institute of Patentees and Inventors: www.invent.org
- ✔ International Trademark Association (INTA): www.inta.org
- ✔ The Motion Picture Association of America: www.mpaa.org
- ✔ The Prince's Trust: www.princes-trust.org.uk
- ✔ Scottish Enterprise: www.scottish-enterprise.com
- ✔ Trevor Baylis Brands plc: www.trevorbaylisbrands.com
- ✔ The UK Trade Association Forum: www.taforum.org
- ✔ Zorin Innovations: www.zorin.co.uk

Mediators and Arbitrators

When dealing with corporations or potential partners you may need to draw on the expertise of organisations versed in dispute resolution. These Web sites can help.

- ✔ The Academy of Experts: www.academy-experts.org
- ✔ ADR Group: www.adrgroup.co.uk
- ✔ The Association of Midlands Mediators: www.ammediators.co.uk
- ✔ CEDR Solve: www.cedrsolve.com
- ✔ The Chartered Institute of Arbitrators: www.arbitrators.org
- ✔ Clerksroom: www.clerksroom.com
- ✔ UK Intellectual Property Office Mediation Service: www.ipo.gov.uk
- ✔ WIPO Arbitration & Mediation Centre: www.arbiter.int

IP Drawing Services

Drawings for patents and designs need to be properly presented. If you're not a professional draughtsman, you might need some help.

- ✔ Appleton Brown Associates Limited: www.appleton-brown.co.uk
- ✔ Garrett & Campbell Limited : www.garrettandcampbell.co.uk
- ✔ Kingfisher Studios: www.kingfisherstudios.co.uk

IP Searching Services

It can be a long slog finding out whether your original idea is truly original: These sites make life easier.

- British Library: www.bl.uk
- RWS Information Limited: www.rws.com
- UK Intellectual Property Office: www.ipo.gov.uk
- Delphion: www.delphion.com
- Patent Seekers Limited: www.patentseekers.com

IP Translation Services

Patenting or otherwise protecting your invention in other countries may mean presenting technical material in a range of languages. If you're not a gifted linguist, you'll need a helping hand.

- ABC Translations : www.abc-translations.co.uk
- HLL Limited : www.hll.co.uk
- RWS Translations Limited: www.rws.com

Index

• *J* •

• *K* •

• *Q* •

• *R* •

• S •

FOR DUMMIES®

Do Anything. Just Add Dummies

HOBBIES

978-0-7645-5232-8

978-0-7645-5395-0

978-0-7645-5476-6

Also available:

Art For Dummies
(978-0-7645-5104-8)
Aromatherapy For Dummies
(978-0-7645-5171-0)
Bridge For Dummies
(978-0-471-92426-5)
Card Games For Dummies
(978-0-7645-9910-1)
Chess For Dummies
(978-0-7645-8404-6)

Improving Your Memory
For Dummies
(978-0-7645-5435-3)
Massage For Dummies
(978-0-7645-5172-7)
Meditation For Dummies
(978-0-471-77774-8)
Photography For Dummies
(978-0-7645-4116-2)
Quilting For Dummies
(978-0-7645-9799-2)

EDUCATION

978-0-7645-5434-6

978-0-7645-5581-7

978-0-7645-5422-3

Also available:

Algebra For Dummies
(978-0-7645-5325-7)
Astronomy For Dummies
(978-0-7645-8465-7)
Buddhism For Dummies
(978-0-7645-5359-2)
Calculus For Dummies
(978-0-7645-2498-1)
Cooking Basics For Dummies
(978-0-7645-7206-7)

Forensics For Dummies
(978-0-7645-5580-0)
Islam For Dummies
(978-0-7645-5503-9)
Philosophy For Dummies
(978-0-7645-5153-6)
Religion For Dummies
(978-0-7645-5264-9)
Trigonometry For Dummies
(978-0-7645-6903-6)

PETS

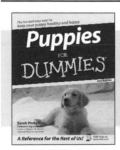

978-0-470-03717-1

978-0-7645-8418-3

978-0-7645-5275-5

Also available:

Labrador Retrievers
For Dummies
(978-0-7645-5281-6)
Aquariums For Dummies
(978-0-7645-5156-7)
Birds For Dummies
(978-0-7645-5139-0)
Dogs For Dummies
(978-0-7645-5274-8)
Ferrets For Dummies
(978-0-7645-5259-5)

Golden Retrievers
For Dummies
(978-0-7645-5267-0)
Horses For Dummies
(978-0-7645-9797-8)
Jack Russell Terriers
For Dummies
(978-0-7645-5268-7)
Puppies Raising & Training
Diary For Dummies
(978-0-7645-0876-9)

FOR DUMMIES®

Do Anything. Just Add Dummies

UK editions

SELF HELP

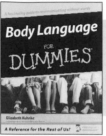
978-0-470-51291-3

978-0-470-03135-3

978-0-470-51501-3

BUSINESS

978-0-7645-7018-6

978-0-7645-7056-8

978-0-7645-7026-1

PERSONAL FINANCE

978-0-7645-7023-0

978-0-470-51510-5

978-0-470-05815-2

Answering Tough Interview
Questions For Dummies
(978-0-470-01903-0)

Being the Best Man
For Dummies
(978-0-470-02657-1)

British History
For Dummies
(978-0-470-03536-8)

Buying a Home on a Budget
For Dummies
(978-0-7645-7035-3)

Buying & Selling a Home For
Dummies
(978-0-7645-7027-8)

Buying a Property in Eastern
Europe For Dummies
(978-0-7645-7047-6)

Cognitive Behavioural Therapy
For Dummies
(978-0-470-01838-5)

Cricket For Dummies
(978-0-470-03454-5)

CVs For Dummies
(978-0-7645-7017-9)

Detox For Dummies
(978-0-470-01908-5)

Diabetes For Dummies
(978-0-470-05810-7)

Divorce For Dummies
(978-0-7645-7030-8)

DJing For Dummies
(978-0-470-03275-6)

eBay.co.uk For Dummies
(978-0-7645-7059-9)

Economics For Dummies
(978-0-470-05795-7)

English Grammar For Dummies
(978-0-470-05752-0)

Gardening For Dummies
(978-0-470-01843-9)

Genealogy Online
For Dummies
(978-0-7645-7061-2)

Green Living For Dummies
(978-0-470-06038-4)

Hypnotherapy For Dummies
(978-0-470-01930-6)

Neuro-linguistic Programming
For Dummies
(978-0-7645-7028-5)

Parenting For Dummies
(978-0-470-02714-1)

Patents, Registered Designs,
Trade Marks and Copyright For
Dummies
(978-0-470-51997-4)

Pregnancy For Dummies
(978-0-7645-7042-1)

Renting out your Property For
Dummies
(978-0-470-02921-3)

Retiring Wealthy For Dummies
(978-0-470-02632-8)

Self Build and Renovation
For Dummies
(978-0-470-02586-4)

Selling For Dummies
(978-0-470-51259-3)

Sorting Out Your Finances
For Dummies
(978-0-7645-7039-1)

Starting a Business on
eBay.co.uk For Dummies
(978-0-470-02666-3)

Starting and Running an Online
Business For Dummies
(978-0-470-05768-1)

The Romans For Dummies
(978-0-470-03077-6)

UK Law and Your Rights
For Dummies
(978-0-470-02796-7)

Writing a Novel & Getting
Published For Dummies
(978-0-470-05910-4)

Available wherever books are sold. For more information or to order direct go to www.wiley.com or call 0800 243407 (Non UK call +44 1243 843296)